THE NORTHERN ADRIATIC ECOSYSTEM

Critical Moments and Perspectives in Earth History and Paleobiology

Critical Moments and Perspectives in Earth History and Paleobiology

David J. Bottjer, Richard K. Bambach, and Hans-Dieter Sues, Editors

Mark A. S. McMenamin and Dianna L. S. McMenamin, *The Emergence of Animals: The Cambrian Breakthrough*

Anthony Hallam, *Phanerozoic Sea-Level Changes*

Douglas H. Erwin, *The Great Paleozoic Crisis: Life and Death in the Permian*

Betsey Dexter Dyer and Robert Alan Obar, *Tracing the History of Eukaryotic Cells: The Enigmatic Smile*

Donald R. Prothero, *The Eocene-Oligocene Transition: Paradise Lost*

George R. McGhee Jr., *The Late Devonian Mass Extinction: The Frasnian/Famennian Crisis*

J. David Archibald, *Dinosaur Extinction and the End of an Era: What the Fossils Say*

Ronald E. Martin, *One Long Experiment: Scale and Process in Earth History*

Judith Totman Parrish, *Interpreting Pre-Quaternary Climate from the Geologic Record*

George R. McGhee Jr., *Theoretical Morphology: The Concept and Its Applications*

Thomas M. Cronin, *Principles of Paleoclimatology*

Andrey Yu. Zhuravlev and Robert Riding, Editors, *The Ecology of the Cambrian Radiation*

Patricia G. Gensel and Dianne Edwards, Editors, *Plants Invade the Land: Evolutionary and Environmental Perspectives*

David J. Bottjer, Walter Etter, James W. Hagadorn, and Carol M. Tang, Editors, *Exceptional Fossil Preservation: A Unique View on the Evolution of Marine Life*

Barry D. Webby, Florentin Paris, Mary L. Droser, and Ian G. Percival, Editors, *The Great Ordovician Biodiversification Event*

The Northern Adriatic Ecosystem

Deep Time in a Shallow Sea

F RANK K. M C K INNEY

 Columbia University Press   *New York*

Columbia University Press
Publishers Since 1893
New York   Chichester, West Sussex

Library of Congress Cataloging-in-Publication Data

McKinney, Frank K. (Frank Kenneth)
    The northern Adriatic ecosystem : deep time in a shallow sea / Frank K. McKinney.
        p. cm. — (Critical moments and perspectives in earth history and paleobiology)
    Includes bibliographical references and index.
    ISBN 978-0-231-13242-8 (cloth : alk. paper)
    1. Paleoecology—Adriatic Sea. 2. Paleoecology—Paleozoic. 3. Marine ecology—Adriatic
Sea. I. Title. II. Series.

    QE720.M42   2007
    577.7'385—dc22   2006038537

This book is dedicated to the biologists and diving team of the Ruđer Bošković Institute Center for Marine Research-Rovinj, who first demonstrated to me the allure of the northern Adriatic; to the Fellows of St John's College, Cambridge, who provided a deeply satisfying and supportive environment in which to start the book; and to Marg, who saw it through from beginning to end.

CONTENTS

ILLUSTRATIONS

TABLES

When I have seen by time's fell hand defaced
The rich proud cost of outworn buried age,
When sometime lofty towers I see down razed,
And brass eternal slave to immortal rage.
When I have seen the hungry ocean gain
Advantage on the kingdom of the shore,
And the firm soil win of the watery main,
Increasing store with loss, and loss with store.
When I have seen such interchange of state,
Or state itself confounded, to decay,
Ruin hath taught me thus to ruminate
That time will come and take my love away.
This thought is as a death which cannot choose
But weep to have, that which it fears to lose.
—SHAKESPEARE, *Sonnet* 64

I have been enamored of the Adriatic Sea and its ecology since I was first taken out in 1987 on a small teaching and research boat, the *Burin*, to sample the bottom-dwelling animals offshore of the picturesque town of Rovinj, Croatia (at that time part of Yugoslavia). I was hoping that there would be a few live colonies of well-calcified, bushy bryozoans that I could study for a better understanding of the ancient bryozoans that I had studied for twenty years. Bryozoans are a group of benthic colonial animals that are one of the most important marine invertebrate *fossil* groups but are considered to be a minor phylum by biologists who study only presently *living* organisms. The scientific literature and European colleagues had given some hope that suitable material might be had off the northeastern Adriatic coast.

On that day in 1987, when the dredge was first hauled up off the seafloor, I was taken totally by surprise and—literally—thrilled by the contents. Here, unceremoniously dumped onto the deck of the *Burin*, was a mound of huge, rigid bryozoan colonies that were filled with writhing brittle stars and crinoids that had used the bryozoans as perches to reach higher into the water for feeding. My mind's eye transformed the whole lot into a mass of mud-covered remains, compacted into rock, and remarkably similar to 450-million-year-old fossil-rich rocks from the Appalachians and Mississippi Valley of North America on which my scientific career had focused until then. That career took a sudden change of direction. Indeed there was a surfeit of material to answer the questions about functional morphology that had brought me to Rovinj. Like so many scientists before, I was—suddenly—passionate about the northern Adriatic. Remarkably, while gently rocking offshore with a picture postcard view of a 450-year-old Venetian town, there at my feet was a pile of animals that, as far as I could see, represented an ecology different from the typical shallow benthic ecology today. It was like peering through a window at animals on a sea floor a million times the age of the 450-year-old town on the shore. What insights about the ancient world might this place yield?

The northern Adriatic Sea has been the focus of studies by marine scientists since the sixteenth century (Zavodnik 1983), and distribution of sediments, benthos, and water mass properties within it are much better known than for most coastal areas of the world. The general similarity of physical and chemical attributes of the northern Adriatic—more precisely the northeastern Adriatic, away from the siliciclastic and nutrient input from the Po River—to Paleozoic epeiric seas further invites examination of its benthic ecosystem. As a paleontologist, I was intrigued to see if it might contribute to a better understanding of the ecology of the Paleozoic fauna.

There have been several books and book-length monographs on various aspects of the Adriatic. These range from Giuseppe Olivi's classic 1792 description of the fauna, *Zoologia Adriatica*, to several late-twentieth-century symposium volumes examining Adriatic sedimentation, marine ecology, and the phenomenon of increasing eutrophication. Cushman-Roisin et al. (2001) have recently reviewed and synthesized the sea's physical oceanography.

This book has a different focus from the others on the Adriatic. Its aim is to make the precarious northern Adriatic ecosystem better known to paleobiologists especially and also to marine ecologists. Jan Seneš was also intrigued by the pertinence of the benthic ecology off the eastern shore of the Adriatic to paleobiology, especially for a better understanding of the marine Cenozoic record exposed in southeastern Europe. He did not produce a single book or book-length monograph but instead published his data and interpretations in a series of papers in the journal *Geologisky Zbornik–Geologica Carpathica* (Seneš 1988a–c, 1989, 1990).

However, having taught oceanography for many years, I cannot jump immediately into a description of the benthos of the Adriatic Sea. In many ways the marine environment is a much more integrated system than is the terrestrial environment; it is impossible to understand the benthos of the sea without at least some rudimentary knowledge of the oceanographic and historical context in which it exists. To this end, all basic oceanographic aspects of the northern Adriatic are important. Their description occupies several chapters. Although the results of some physical models are presented here, elucidation and discussion of the theories behind the models and other conceptual aspects of basic oceanography are outside the scope of this study and are not covered here except tangentially, where necessary for understanding the description given.

Chapter 1 of the book describes the fundamental evolutionary ecological conundrum for which I think the northern Adriatic is particularly informative as a living laboratory. The next five relatively brief chapters of the book give the context of continual change in which the present northern Adriatic exists: general physical geography; geological origin; water masses involved, their physical characteristics and circulation; distribution and concentration of nutrients within the water column and the pelagic ecosystem; and accumulation and type of sedimentary fill in the basin. The subsequent chapters examine the benthic ecosystem, its relationship to nutrients within the water column and to sediments, its response to environmental fluctuations and stresses, and preservation potential of the living benthos. In the final chapter I discuss the pertinence of the northern Adriatic ecosystem to the macroevolution of marine benthic ecology during the Phanerozoic. An epilogue summarizes the current human-generated environmental pressures that have altered or threaten to alter the ecology of this rather small body of water, and the geological future of the entire Adriatic. Overall, it is my intention to show that a broader understanding of the northern Adriatic Sea will give increased insight into the benthic ecosystems of ancient shallow seas and their ecological evolution.

This book could not have been written without a wide range of support from an enormous number of individuals and institutions that deserve thanks. Exposure to the rich northern Adriatic fauna and environment was provided by grants from the U.S. National Academy of Science–Yugoslav Academy of Sciences exchange program, the National Geographic Society, and Appalachian State University. The entire staff of the Ruđer Bošković Center for Marine Research–Rovinj was amazingly cooperative and helpful, but two people there must be singled out because I depended on them so very much: Dušan Zavodnik and Andrej Jaklin. Simon Conway Morris invited me to apply to spend a term as Visiting Scholar at St. John's College, Cambridge, where the bulk of library work and compilation of notes for the book was completed, in what must be one of the most congenial and supportive academic settings in the world. But even the huge library of the University of Cambridge doesn't hold each and every limited-circulation regional journal, and librarians Libby Tilley at University of Cambridge's Earth Sciences library and Dianna Moody Johnson at the library of Appalachian State University tracked down an untold number of out-of-the-way papers that I stumbled onto. Allen Smith and Deborah Howard, both of St John's College, enlarged my perspective of the Adriatic, and innumerable Italian and Croatian scientists generously shared the results of their research. Katherine Graham and coworkers couldn't make me level-headed, but did keep me on my feet through the time at St John's. I also owe sincere thanks to colleagues who vetted chapters of the book for the most egregious errors: Danilo Degobbis, Ole Gade, Steve Hageman, Patricia Kelley, and Loren Raymond. Steve Hageman helped immensely by intermittent tutoring on various statistical analyses.

My most profound thanks, however, go to Marg, who has been my support throughout, at the beginning figuratively and increasingly toward the end, literally.

THE NORTHERN ADRIATIC ECOSYSTEM

CHAPTER ONE

LONG-TERM CHANGES IN SHALLOW MARINE LIFE

Ships that pass in the night, and speak to each other in passing,
Only a signal shewn, and a distant voice in the darkness.
So on the ocean of life we pass and speak to one another.
Only a look and a voice; then darkness again and a silence.
—HENRY WADSWORTH LONGFELLOW, *Tales of a Wayside Inn*

Ah, Peter, we are but ships that go bang in the night.
—BBC production of *A Year in Provence*

EARLY DIVERSIFICATION OF MARINE ANIMALS

A little over 540 million years ago an important transformation in the
marine ecosystem was in progress. About then, and during the next
approximately 40 million years, bottom-dwelling invertebrate animals
with mineralized skeletons appeared and became increasingly diverse
and abundant. Sponges, corals, mollusks, brachiopods, echinoderms, and
arthropods with mineralized skeletons spread into diverse shallow marine environments around the world.

The same process of rapid exploration of body plans and diversification
was apparently happening at the same time among "soft-bodied" animals,
some of which later developed lineages with mineralized skeletons. Representatives of these groups are found around the world in several outcrops of
fine-grained sediment deposited in anoxic regions of these ancient seas.
These Burgess Shale–type deposits have yielded impressions of soft-bodied
animals such as annelids, priapulids, ascidians, and chordates, along with
many organisms that do not fit within the well-defined phyla recognized by
students of today's marine life.

Therefore, within the interval of time known as Early Cambrian, the
phyla of living benthic marine animals were established, along with their
fundamental ecological roles that we see today: suspension-feeders, detritus-feeders, deposit-feeding sediment-swallowers, carnivores, and—in

shallow water—herbivores. (There is one startling absence throughout the Cambrian: the Bryozoa. This phylum has a rich fossil record of well-mineralized skeletons in younger rocks but has not been found at all in Cambrian rocks.) Some of the Early Cambrian animals lived in the water above the seafloor, others lived on the surface of the rocks and sediments of the seafloor, and preserved burrows indicate that others lived within the sediments. In a sense, the basic organization of today's benthic marine ecosystem was established along with the diversification of Early Cambrian animals. However, important changes in dominance of one group relative to another have occurred during the past 540 million years, with interesting ecological consequences.

Faunal Succession

None of the species that were part of the Early Cambrian marine fauna are still alive. Although the living marine fauna is made up of distantly evolved descendants of Cambrian animals, they still have the same fundamental body plans and limitations on sources of nutrients. Origination and extinction of millions of species has occurred since the advent of the Cambrian so that now, 540 million years later, there has been a long succession of different faunas that have occupied the world's oceans.

Marine invertebrate species in general have existed over intervals of approximately 2 to 8 million years, dependent in part on whether their geographic distribution was relatively local or more cosmopolitan (reviewed in Jablonski 2000). If marine invertebrate species on average have existed for about 5 million years, and the timing of evolution of new species has been completely random, then for every 5-million-year increment of earth history the seas would have experienced an almost complete faunal turnover. A minority of species would have come and gone completely within that 5-million-year interval, and a minority that were alive 5 million years previously would still be living at the end of a 5-million-year interval. However, most of the species alive at the beginning of the interval would have become extinct, to be replaced by an approximately equal number of newly evolved species.

If there has been on average a complete faunal turnover every 5 million years, then there has been a temporal series of about 108 different faunas. Paleontologists that are particularly good at recognizing fine-scale differences between species can discriminate even more subdivisions of time. This can be demonstrated for a sequence of marine sedimentary rocks within the British Isles that represents only about three-quarters of the past 540 million years. But within that partially complete regional sequence, 216 successive faunal zones can be tallied (Duff and Smith 1992).

The sequence of faunal zones within stacked sedimentary rocks began to catch the eye of paleontologists in the late eighteenth century, most notably George Cuvier in France and William Smith in England. During the later decades of the eighteenth century and the first half of the nineteenth century, the sequence of sedimentary rocks was the focus of intense intellectual engagement and debate. Discussion of the intellectual ferment and conflict during this time is not part of this book, but numerous interesting books on the topic were written during the late twentieth century by paleontologists, historians of science, and others (e.g., Berry 1968; Rudwick 1972, 1985; Secord 1986; Winchester 2001).

During the late eighteenth and early nineteenth centuries, the sedimentary sequence was divided into the rock systems and time periods (fig. 1.1) that are the core of the terminology and concepts that geologists use when talking about time: Cambrian, Silurian, Devonian, Carboniferous, Permian, Triassic, Jurassic, Cretaceous, Tertiary, Quaternary, all lying ultimately on even older rocks all collectively and informally known as "Pre-Cambrian." The Ordovician System was named later, applied to a disputed sequence of overlap between the originally defined Cambrian and Silurian Systems. Rather than refer to Tertiary and Quaternary, many geologists now prefer to use the terms Paleogene and Neogene, which were introduced later and subdivide the most recent part of the sedimentary sequence a bit differently. Each of the systems of rock was distinguished by its position in relation to underlying and overlying rocks, but each was also recognized to contain a unique suite of fossils by which rocks formed during the same time could be recognized elsewhere.

The time during which Cambrian through Quaternary rocks formed is referred to collectively as the Phanerozoic because of the readily visible marine and eventually terrestrial fossils found in them. Adam Sedgwick coined the term "Palaeozoic" (i.e., ancient life) in 1838 in a talk for the Geological Society of London as a collective term for Cambrian and Silurian rocks. Shortly thereafter John Phillips (1840) suggested that the term be extended for the British stratigraphic sequence from the Cambrian through Magnesian Limestone. At the time the Permian System, which is the uppermost part of the Paleozoic sequence, had not yet been recognized but was named soon thereafter based on rocks in Russia that formed at the same time as the Magnesian Limestone. Phillips characterized Paleozoic rocks—and the time interval that they represent—by combinations of various tabulate corals, brachiopods, cephalopods, and trilobite fossils within them. In the same paper, Phillips first mooted the terms "Mesozoic" and "Kainozoic" (now referred to as Cenozoic) without defining them, but then defined them the following year (Phillips 1841) essentially as they are used today.

Geologists keep in mind the difference between the rocks that are seen on the ground and the time during which they formed. We should be able

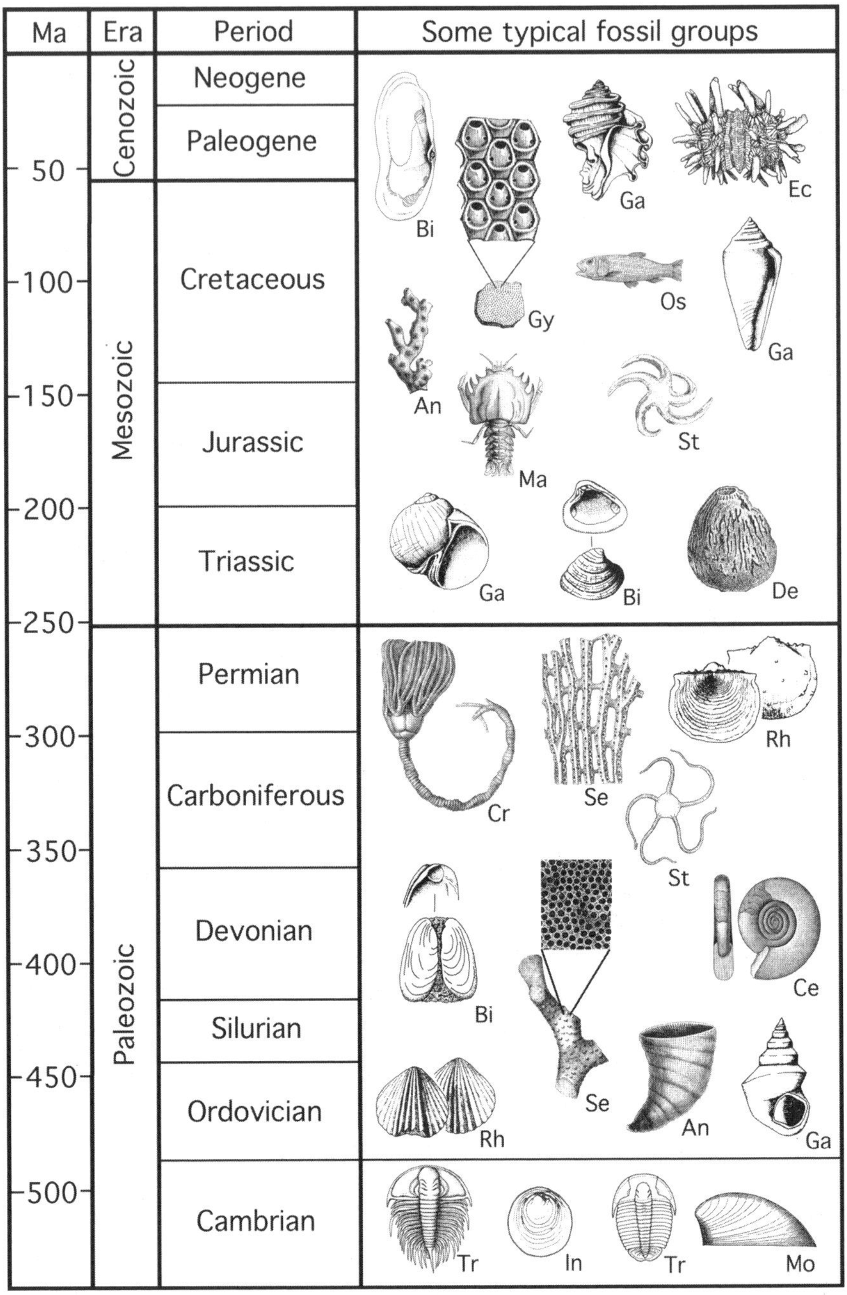

FIGURE 1.1 The Phanerozoic time scale with representatives of most of the characteristic classes of skeletalized marine animal fossils.

Key to classes: An = Anthozoa, Bi = Bivalvia, Ce = Cephalopoda, Cr = Crinoidea, De = Demospongia, Ec = Echinoidea, Ga = Gastropoda, Gy = Gymnolaemata, In = "Inarticulata" (Linguliiformea and Craniformea), Ma = Malacostraca, Mo = Monoplacophora, Os = Osteichthyes, Rh = "Articulata" (Rhynchonelliformea), Se = Stenolaemata, St = Stelleroidea, Tr = Trilobita.

Time scale calibrated according to Gradstein et al. (2004); fossil thumbnail drawings from McKinney and McKinney (1991), Nicholson (1872), and Ulrich (1890).

to match up rocks of equivalent age that formed anywhere in the world during the past 540 million years with rocks that constitute the original sequence of Cambrian through Neogene rocks in England and Europe. The match-up can be based on similar fossil content, radiometric age-dating, or either of these used in combination with several other chemical or physical techniques. We can thus discuss the rocks and the fossils in them either as physical objects, or as representatives of a unit of past time. There are parallel sets of terms for the sequence of rocks and for the sequence of time (table 1.1). For example, if we talk about Lower Silurian rocks, they belong to the Llandovery Series of the larger Silurian System, which is part of the Paleozoic Erathem. Each of these progressively larger sets of rocks formed during intervals of time that are called the Early Silurian—or Llandovery—Epoch of the larger Silurian Period, which is part of the Paleozoic era.

To be precise, geologists should specify whether the topic is the Silurian *System* or the Silurian *Period*. However, it is tedious to repeat "System" or "Period" after each use of Silurian, so we often drop the time/rock specifier and hope that it will be understood from context whether time or rock is meant. A key to whether rocks or time is indicated is the use of the terms "Lower" and "Upper" (indicative of rocks and the fossils in them) or "Early" and "Late" (indicative of time or of the animals and plants as they were alive at that time).

The organization of Phanerozoic time into periods, and its larger three-fold subdivision into eras based on large-scale faunal changes, has been fully supported by subsequent work around the world. The eras are divided one from the other by the two largest extinctions since the beginning of the Cambrian (Raup and Sepkoski 1982), and other large extinctions correspond exactly or nearly with other system boundaries. But not all system boundaries are marked by conspicuous, larger-than-background extinctions.

TABLE 1.1   Parallel Sets of Terms Used to Denote Rock Units and Time Units

TIME UNITS	ROCK UNITS
Era	Erathem
Period	System
Epoch	Series
Age	Stage

Time units: terms defining specified portions of geological time and arranged hierarchically from the largest at the top of the table to the smallest at the bottom.

Rock units: terms applied to all rocks that formed on Earth during a portion of time that received a two-word name that ends with the term in the column to the left.

In the late twentieth century Jack Sepkoski published a series of remarkable papers that summarized and analyzed the Phanerozoic patterns of diversity, origination, and extinction of marine animals (1978, 1979, 1981, 1982, 1984). He examined a huge range of paleontological literature to determine the stratigraphic ranges of genera, but confined his analyses to family and higher-level taxa. This naturally generated a prolonged discussion of whether the traditional, qualitative definitions of higher taxa have any useful validity for quantitative studies. Nonetheless, the basic patterns of origination, diversification, and extinction of taxa that Sepkoski found have held up to further testing.

Sepkoski (1981) used factor analysis to look at number and association of marine families within metazoan animals through the Phanerozoic. The first three factors show strong associations of largely different taxonomic groups that characterize three successively dominant marine faunas (fig. 1.2). The analysis confirms that post-Paleozoic marine faunas differ greatly from earlier faunas. But rather than discrimination between Mesozoic and Cenozoic faunas, it turns out that there is little difference between them in the higher-level taxonomic groups. Instead, there was a

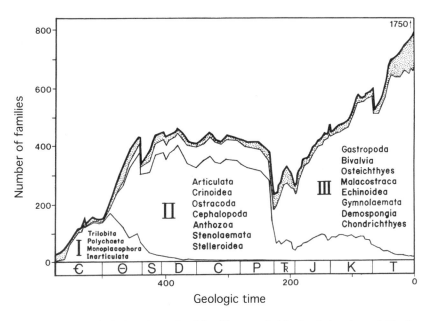

FIGURE 1.2 Number of marine animal families recorded in Cambrian through Tertiary rock, organized into the three faunas determined by factor analysis: Cambrian (I), Paleozoic (II), and Modern (III). Shaded area along the top of the diversity curve represents diversity of soft-bodied fossils and is not included in any of the three successive faunas.

From Sepkoski 1981.

profound difference between Cambrian and later Paleozoic faunas, which is qualitatively shown in figure 1.1. The uniqueness of the Cambrian fauna had been noted from the time that the System was first named, but Sepkoski's factor analysis indicated a greater contrast between composition of the Cambrian and the later marine Paleozoic faunas than exists between Mesozoic and Cenozoic faunas. Sepkoski named these the Cambrian, Paleozoic, and Modern evolutionary faunas, indicated by Roman numerals I to III in figure 1.2.

As can be seen in figure 1.2, the origin of all three faunas is to be found at about the Precambrian–Cambrian boundary. However, the classes within the three successive evolutionary faunas diversified exponentially in three successive phases: Early Cambrian for the Cambrian fauna, Early Ordovician for the Paleozoic fauna, and a more extended diversification for the Modern fauna. The Modern fauna was moderately well developed by the Carboniferous as the Paleozoic fauna was slightly declining. However, the Modern fauna became predominant only after the end-Permian extinction, while the Paleozoic fauna never recovered vigorously after the end-Permian extinction and has declined substantially starting in the Late Cretaceous.

ECOSYSTEM SUCCESSION

The late 1970s and early 1980s was a time of vigorous intellectual ferment within paleoecology. Much of that conceptual development was elaborated in a book, *Biotic Interactions in Recent and Fossil Benthic Communities*, edited by Tevesz and McCall (1983). Chapters in that book by Richard Bambach, Charles Thayer, and Geerat Vermeij are particularly germane in setting the context for the northern Adriatic benthos's importance in understanding ecological changes through the Phanerozoic.

Richard Bambach's paper in Tevesz and McCall's book was entitled "Ecospace utilization and guilds in marine communities through the Phanerozoic" and addressed Sepkoski's three evolutionary marine faunas as an ecological theater: "Do these three groupings of taxa differ in some way that reflects their biological or ecological response to the world? The major adaptive themes of the dominant organisms in the three evolutionary faunas should indicate the degree of ecospace exploitation by each fauna. If more ecospace is exploited by more recent faunas, then increased complexity in the utilization of the ecosystem has accompanied the change from dominance by one fauna to another. An ecologic component would be identified in the course of evolution" (Bambach 1983:721–722).

Bambach (1983, 1985) focused on the fossil record of shelf-depth marine faunas, including remains both from the flooded margins of continents and from shallow, epeiric seas that spread across broader stretches of continents. His paper characterized each of the three evolutionary faunas by the dominant life habits within its characteristic taxa. He defined the various life habits by the organisms' relation to the sediment–water interface, whether mobile or sedentary, and by their principal feeding type:

I. Relationship to sediment–water interface
  A. Pelagic
  B. Epibenthic
    1. Mobile
    2. Attached
      a. Low
      b. Erect
    3. Reclining
  C. Endobenthic
    1. Shallow
      a. Passive
      b. Active
    2. Deep
      a. Passive
      b. Active
II. Principal feeding method
  A. Suspension-feeder
  B. Detritus-feeder
  C. Deposit-feeder
  D. Herbivore
  E. Grazing omnivore
  F. Carnivore

Bambach considered life habit (position/mobility X feeding type) in combination with morphological groups within a taxonomic class to constitute a broadly defined guild. (This use of the term "guild" is at the most broadly defined end of the spectrum, because it is typically used for smaller taxonomic groupings functioning within more finely divided ecospace.) The life habits as defined by Bambach have been termed "Bambachian megaguilds" (Droser et al. 1997). This is the ecological level at which benthic organisms most commonly are grouped in this book, and the resulting ecological groups will be referred to as megaguilds. Because of indeterminate boundaries or simply for convenience in assigning species, some groups are combined for analysis in chapters 7 through 9: the endobenthos is not subdivided into shallow versus deep, and grazing omnivores and herbivores are treated as a single group.

CAMBRIAN BENTHIC ECOSYSTEM

The Cambrian fauna was dominated by epibenthic organisms (fig. 1.3), the most abundant of which had shells of calcium phosphate. Trilobites are the most common Cambrian mineralized fossils in sediments deposited at shelf depths, followed by "inarticulate" brachiopods, hyoliths, monoplacophorans, and relatively few but diverse other shelled invertebrates. Trilobites were epibenthic, crawling and swimming detritus-feeders and grazers, and hyoliths and monoplacophorans also appear to have been mobile detritus-feeders. Inarticulate brachiopods were epibenthic to partially buried sedentary and were suspension-feeders; most of the less common shelled invertebrates were epibenthic, sedentary suspension-feeders. Shallow burrows indicate that some inarticulate brachiopods and soft-bodied organisms such as polychaetes penetrated into the upper few centimeters of sediment.

PALEOZOIC BENTHIC ECOSYSTEM

The Paleozoic fauna was more diverse and generally more abundantly preserved (fig. 1.4), owing largely to the rapid Ordovician proliferation of taxa with skeletons of calcium carbonate. Epibenthic, sedentary suspension-feeders dominated most associations. Many fed from just above the sediment–water interface, including attached and free-lying "articulate" brachiopods; byssally attached bivalves; free-lying bryozoans and corals; and diverse encrusting animals including sponges, rugose and tabulate corals, and stenolaemate bryozoans. Many others fed well above the sediment–water interface, including crinoids and other stalked echinoderms, and erect-growing sponges, tabulate corals, and stenolaemate bryozoans, as well as any animals that normally fed just above the sediment–water interface but whose spat happened to attach to the taller animals and algae. Diverse mobile epibenthos included deposit-feeders (monoplacophorans, gastropods), herbivores and grazers (echinoids, gastropods, monoplacophorans, crustaceans), and carnivores (cephalopods, crustaceans, starfish), as well as swimming mineralized carnivores (cephalopods, chondrichthyes, osteichthyes).

From Ordovician through Permian, sediment-penetrating organisms were secondary in most environments but were more abundant and diverse than during the Cambrian. Some suspension-feeding bivalves inhabited shallow stationary burrows while others moved through the upper few millimeters of sediment while either suspension feeding or deposit feeding. Deposit-feeding polychaetes along with a few trilobites and carnivorous

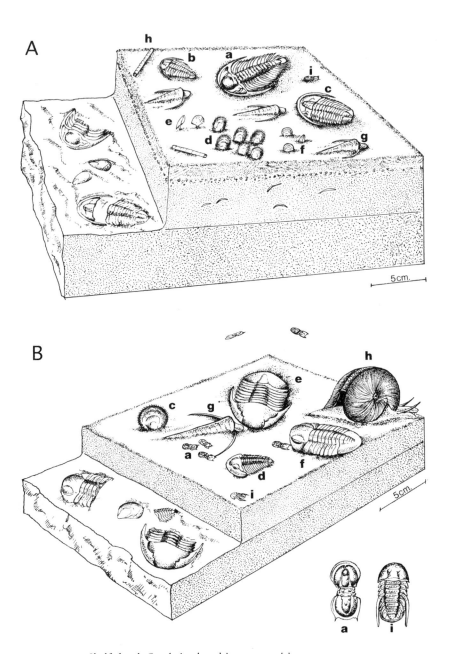

FIGURE 1.3 Shelf-depth Cambrian benthic communities.

A) Shallow water shelly community consisting of trilobites (a, b, c, i), inarticulate brachiopods (d, f), articulate brachiopods (e), hyoliths (g), and tubular remains of uncertain affinities (h).

B) Deeper shelf shelly community of trilobites (a, d, e, f, i), inarticulate brachiopods (c), hyoliths (g), and monoplacophorans (h).

From McKerrow 1978.

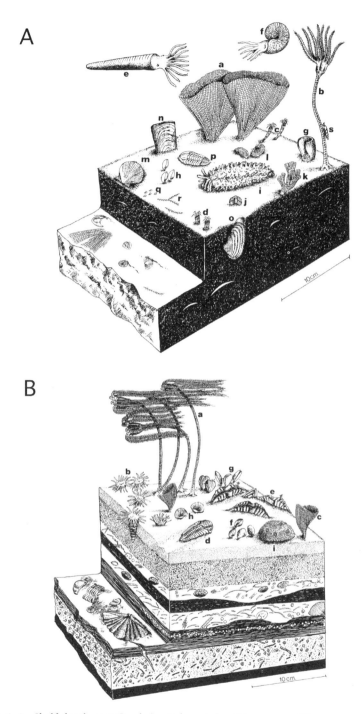

FIGURE 1.4   Shelf-depth post-Cambrian Paleozoic benthic communities.

A) Carboniferous mud-dwelling community consisting of stenolaemate bryozoans (a, k, r), crinoids (b), corals (c, d), cephalopods (e, f), articulate brachiopods (g, h, j, l), holothurian (i), bivalves (m, n, o), trilobites (p), ostracodes (q), and gastropods (s).
B) Devonian mud-dwelling community consisting of crinoids (a), corals (b, i), stenolaemate bryozoans (c), trilobites (d), and articulate brachiopods (e, f, g, h).

From McKerrow 1978.

polychaetes also disturbed the shallow sediment, but only a few deposit-feeding bivalves penetrated more deeply into the sediment.

With the end-Permian extinction, many of the dominant epibenthic Paleozoic clades were completely eliminated or so severely reduced that they either drifted to eventual extinction or have never recovered vigorously. Some of the major groups that went extinct at the end of the Permian include the trilobites, most major clades of brachiopods, the two clades of Paleozoic corals, and the three Paleozoic clades of crinoids (although one was paraphyletic, giving rise to the post-Paleozoic crinoids). All but the most minor group of the Paleozoic stenolaemate bryozoans either were abruptly terminated by the end-Permian extinction or were severely reduced and became extinct within a few tens of millions of years.

Modern Benthic Ecosystem

Groups that survived the end-Permian extinction relatively unscathed diversified vigorously within oceans that were undersaturated with respect to taxonomic richness (fig. 1.2). As a consequence the marine ecosystem experienced a major reorganization. Some bivalves remained as part of the epibenthos, but many suspension-feeding lineages diversified as endobenthos. Malacostracans also diversified vigorously, many seeking refuge from predators by moving into the sediment, either excavating permanent burrows or digging a series of temporary shelters. A major group of echinoids developed biradial symmetry as they moved into the sediment and became detritus-feeders. Simultaneously, several diversifying groups gave rise to progressively more effective predators scouring the seafloor and even probing into the sediment in search of prey: gastropods, cephalopods, malacostracans, radial echinoids, osteichthyes, and asteroids.

By late Mesozoic, as at present, the preservable shallow marine sediment-dwelling benthos was dominated by endobenthic bivalves, sediment-probing gastropods, endobenthic and epibenthic echinoids, mobile malacostracans, and in some environments by rapidly growing scleractinian corals or by rooted upright or free-living cheilostome bryozoans. Sedentary, epibenthic suspension-feeders were and are rare to absent on accumulations of fine marine sediment, except on small-scale, cryptic surfaces of skeletal debris lying on the sediment surface (e.g., Lescinsky 1993; McKinney 1996; Nebelsick et al. 1997; Zuschin et al. 1999).

The Modern fauna has a rather different ecological structure from the Paleozoic fauna. It is dominated by endobenthos and mobile epibenthos but has few if any sedentary epibenthic suspension-feeders (fig. 1.5). Many suspension-feeding, deposit-feeding, and carnivorous organisms now churn through the uppermost 30 cm or so of sediment, including bivalves,

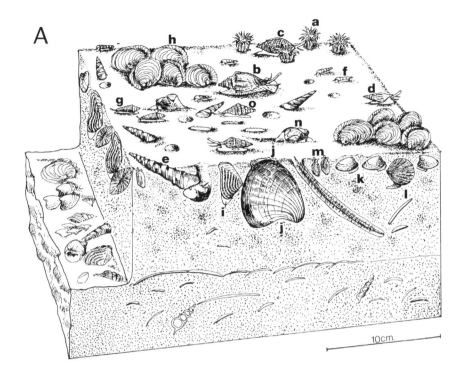

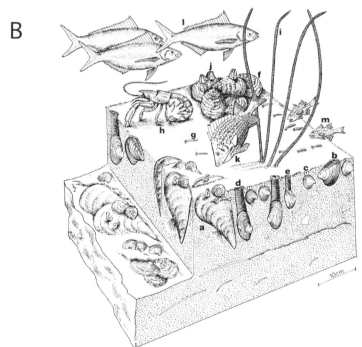

FIGURE 1.5 Shelf-depth Paleogene benthic communities.

A) Eocene silt/sand community consisting of corals (a), gastropods (b, c, d, e, f, g, n, o), bivalves (h, i, j, k, l, m), and polychaetes (occupying burrow).

B) Eocene sandy clay community consisting of bivalves (a, b, c, d, e, f), annelids (g), malacostracans (h), octocorals (i), barnacles (j), and osteichthyes (k, l, m).

From McKerrow 1978.

polychaetes, echinoids, holothurians, gastropods, and malacostracans. And a surprising diversity with the same range of feeding has penetrated deeply (to a meter or more) into the sediment, including bivalves, polychaetes, echinoids, and malacostracans.

INVASION OF MARINE SEDIMENT BY ANIMALS

The general pattern through the Phanerozoic, therefore, has been that potential marine benthic ecospace associated with sedimentary bottoms has been progressively occupied more extensively through time by animals (fig. 1.6; Bambach 1983, 1985). The Cambrian fauna was dominated by a limited number of epibenthic life habits: attached low suspension-feeders, deposit-feeding and grazing mobile epibenthos; and only a few organisms had penetrated slightly into the sediment. During the post-Cambrian Paleozoic all the epibenthic life habits were occupied by diverse, abundant animals. Although a greater range of endobenthic life habits had come into existence, they were not intensely occupied during the Paleozoic, with the exception of siliciclastic environments where mollusks were prominent (Miller 1988, 1989). The rise of the Modern fauna (foreshadowed by the mollusk-dominated assem-

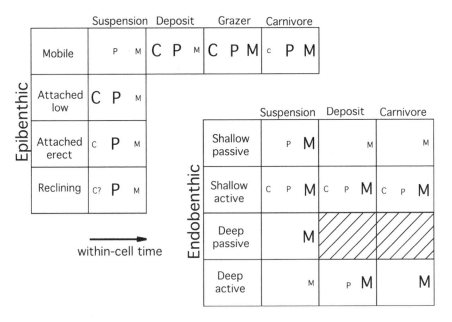

FIGURE 1.6 Benthic marine megaguild distribution for Cambrian (C), Paleozoic (P), and Modern (M) faunas, based on information from Bambach 1983. Smaller letters indicate that the megaguild was present but not a major part of the biomass on sedimentary substrata. Note the progressive shift in dominance from epibenthic megaguilds (upper left block of cells) to endobenthic megaguilds (lower right block of cells).

blages during the Paleozoic) shifted primary ecospace occupancy into the sediment, including many actively moving animals.

Bambach (1983) found an average of 11 broadly defined guilds in local Paleozoic faunas but an average of about 18 in local Neogene faunas, which also were more species-rich than the Paleozoic faunas. However, even for comparable local faunas with 50–70 species (the maximum for Paleozoic faunas studied by Bambach), there were more guilds in the Neogene faunas. The greater utilization of ecospace by the Modern fauna is largely the result of diverse taxa invading and living within the sediment rather than the result of niches within epibenthic guilds being more finely divided (Bambach 1983).

The patterns summarized above are only a broad overview of more detailed patterns based on very large data sets collected and analyzed by— among others—Ausich and Bottjer (1982), Bambach (1983, 1985), Thayer (1983), and Sepkoski and Miller (1985). However, the differences in softsubstrate benthic ecosystems through time summarized in the past several paragraphs are seen by anyone who has collected fossils from rocks distributed across the Phanerozoic column.

In fact, figures 1.3–1.5 are reproduced from *The Ecology of Fossils*, a book edited by W.W. McKerrow and published in 1978. This book is based on descriptions of 119 well-preserved representative Cambrian through Neogene fossil assemblages plus six presently living shallow marine communities, most found in the British Isles. It includes a full-page block diagram drawing of each assemblage, intended to represent what the various marine benthic communities looked like in life, along with a strip on one side of the block diagram that shows the nature of the skeletal remains from which the concept of the association was based. The six drawings that constitute figures 1.3–1.5 are drawings of shelf-depth mud to fine-sand benthic communities, from each of the three evolutionary faunas. *The Ecology of Fossils* was published before Jack Sepkoski's papers in which the three evolutionary faunas were characterized by factor analysis, and before Richard Bambach analyzed their basic ecological structure. The entire suite of illustrations in *The Ecology of Fossils* seems a prescient confirmation of both Sepkoski's and Bambach's analyses.

HYPOTHESES ABOUT CAUSES OF LONG-TERM ECOLOGICAL CHANGES

The important change from a shelf-depth benthic ecosystem anchored in epibenthic suspension-feeders to one dominated by endobenthos wasn't a necessary outcome of the end-Permian marine extinction. Many lineages of epibenthic suspension-feeders survived this most-intense Phanerozoic extinction event. Representatives of crinoids, all three sponge classes, corals, almost all major clades of stenolaemate bryozoans, several clades of articulate brachiopods, and epibenthic attached and free-lying bivalves all survived the end of the Permian. Yet most stenolaemate bryozoan clades were extinct well before the end of the Mesozoic. Crinoids and articulate brachiopods are virtually restricted to cryptic or deep-water environments. Among these formerly abundant epibenthic suspension-feeders, only sponges and corals are widespread and locally common in exposed shallow-water environments.

Why the change in the structure of the benthic ecosystem from Paleozoic to Modern faunas? Despite the shift in biomass from above the sediment–water interface into the sediment, there has been only a small increase from the Paleozoic in the range of benthic marine life habits (fig. 1.6). Even if sedentary epibenthic suspension-feeders were preferentially affected by the end-Permian extinction, survivors in diverse lineages should simply have had the opportunity to exploit the absence of competitors and to diversify rapidly. Their failure to move vigorously back into and to hold the relatively empty epibenthic ecospace across the range of shallow-water environments suggests that something about benthic ecology had changed or was changing.

Various long-term environmental changes have been suggested as possible driving forces for replacement of the largely epibenthic Paleozoic ecology by the endobenthos-rich Modern ecology. These include (1) orders of magnitude increase in bioturbation, (2) substantial diversification of predators and increase in intensity of predation, and (3) increased nutrient concentration in coastal marine waters.

Sediment Flux and Bioturbation

Epibenthic attached suspension-feeders are subject to burial by high rates of sedimentation. In addition, local disturbance—resuspension and local resettlement—can cause a high rate of sediment flux across the sediment–water interface. Suspended sediments that clog the feeding apparatus adversely affect all suspension-feeders, including endobenthos.

For example, mobile suspension-feeding ophiuroids (Schäfer 1972) as well as sedentary epibenthic bryozoans are precluded by high rates of sediment flux. Sediment accumulation approaches 40 cm yr^{-1} off the Rhône Delta on the south coast of France. Within this region bryozoans are entirely absent, and even mobile endobenthos are reduced in species richness and number (Lagaaij and Gautier 1965).

Sediment instability is increased by any organism that disturbs the sediment–water interface (table 1.2), including especially all mobile endobenthos and epibenthic sediment-probing organisms, collectively known as bioturbators. At present several groups of epibenthic predators probe into

TABLE 1.2   Types of Sediment–Mediated Biotic Disturbances

Process	"Villains"	"Victims"
Bulldozing		
Displacement	Any mobile organism	Immobile (exhumed, buried, disoriented)
Manipulation (while burrowing or crawling)	Any mobile organism	Immobile (exhumed, buried, disoriented)
Manipulation (during feeding)	Epibenthic predators and scavengers of infauna	Immobile (exhumed, buried, disoriented)
Ingestion and egestion (while sediment mining)	Deposit-feeders	Immobile
Pseudofeces (from deposit feeders)	Deposit-feeders	Immobile
Biodeposition (defecation and pseudofeces)	Suspension-feeders	Immobile, especially small suspension-feeders
Biotic resuspension (overlaps bulldozing in part)	Deposit-feeders in muds	Suspension-feeders (by clogging of feeding apparatus)
Fluidizing of sediments	Deposit-feeders in muds, possibly other mud disturbers	Immobile epibenthos (sink into fluidized sediment)

"Villains" = perpetrators, and their habits
"Victims" = animals affected, and their habits
After Thayer 1983.

the sediment for prey. Many crabs and gastropods can probe to a depth of several centimeters for bivalve prey (Stanley 1985). Several species of sea stars grub bivalve prey out of sand to depths of at least 40 cm (e.g., Sloan and Robinson 1983). The cow-nosed ray, *Rhinoptera bonasus*, can plow depressions up to a meter wide and to a 45 cm–depth in search of bivalves (Orth 1975). These are just a few examples of sediment disturbance by epibenthic and pelagic predators that probe down into sediment from the sediment–water interface (see reviews in Thayer 1983 and Vermeij 1987).

Epibenthic organisms other than sediment-probing carnivores also contribute to instability of the sediment–water interface and add to the hazard of burial for any epibenthic sedentary organism. Many crabs scuttle backwards or sideways into the substrate so that they are covered by a thin, camouflaging layer of sediment; in the process they may tip over and/or push sediment onto sedentary organisms. Mobile suspension- and detritus-feeding holothurians can drop their fecal pellets or push sediment onto low-profile sedentary animals or plants. Some suspension-feeding epibenthos and partially sediment-immersed benthos ingest large volumes of sediment from water passing by, and that sediment is then passed as bundles or strings loosely bound by mucous, known as pseudofeces. For example, the marsh bivalve *Geukensia demissa* feeding in sediment-rich New England waters annually produce such a volume of pseudofeces that they deposit approximately the same amount of particulate nitrogen as is tidally flushed from the marshes (Jordan and Valiela 1982). Where *G. demissa* are common, sediment floors of marshes may be completely covered by semicoherent pseudofeces.

Endobenthic burrowers shift even more sediment. Amount of sediment disturbed by endobenthic animals is determined by their density (number per unit area), cross-sectional size of the burrow generated, and rate at which burrowing occurs. Animals that live within the sediment can occupy stationary burrows that may slowly extend as the animal needs more space in which to live, or they may actively move through the sediment. Many but not all endobenthic suspension-feeders and detritus-feeders that feed at the sediment–water interface occupy stationary burrows. Movement of mobile endobenthos may be relatively slow; many mobile endobenthic suspension-feeders commonly move slowly, except that some can burrow rapidly back into the sediment when threatened by predators or after being exposed. Endobenthic deposit-feeders and carnivores commonly burrow more rapidly and can disturb large volumes of sediment. In the Wadden Sea, populations of the lugworm *Arenicola marina* annually can rework sediment volumes equal to the entire upper 33 cm depth (Cadée 1976).

A rapid increase in bioturbation during the Mesozoic is hypothesized to be a direct cause of the decrease in sedentary epibenthic suspension-feeders (Thayer 1979, 1983). The increase in bioturbation during the Mesozoic resulted from increases in diversity, volume shifted by individual bioturbators, and total population reworking rate.

Figure 1.7 includes a plot of total diversity through the Phanerozoic fossil record of extant bioturbating groups. Taxonomic level varies from one "group" to another; some are families and others phyla. Subjectivity in tracking the record of bioturbating groups is required because the intensity of bioturbation of certain taxonomic groups within phyla warrants noting their individual origin, such as the advent of clam-digging sea stars early in the Mesozoic, whereas the Asterozoa originated well back in the Ordovician, and the phylum Echinodermata, to which the asterozoans belong,

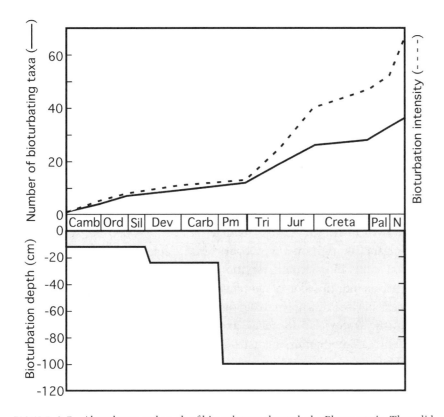

FIGURE 1.7 Abundance and reach of bioturbators through the Phanerozoic. The solid line in the upper part of the figure tracks the cumulative number of extant bioturbating groups as determined from their oldest known fossils, compiled from information in Thayer (1983). The dashed line gives a rough, conservative impression of the increase in bioturbation intensity through time. It was fabricated by the expedient of multiplying each existing bioturbating lineage by a factor of 0 to 4 depending upon the $\log_{10}$ of the group's sediment-reworking rate in $cm^3\ day^{-1}$ ($0 = <10^{-1}$, $1 = 10^{-1}–10^1$, $2 = 10^1–10^3$, $3 = 10^3–10^5$, $4 = >10^5$) as given in Thayer 1983. Departure of the two lines in the upper part of the figure from the Triassic to the Recent reflects the post-Paleozoic addition of organisms that disturb sediment at a more rapid rate than almost all in the Paleozoic. The lower part of the figure shows depth of bioturbation as determined from the rock record (after Ausich and Bottjer 1982; Bottjer and Ausich 1986).

originated early in the Cambrian. Burrowing echinoids can be traced back to the Jurassic and constitute a second important group of echinoderms that evolved a bioturbating habit and so should be recorded independently. On the other hand, the phylum Priapula in general are sediment-dwellers and most are active burrowers, so in contrast with the phylum Echinodermata, Priapula are scored as a bioturbating group from their earliest appearance in the Cambrian.

The number of bioturbating groups more than doubled from the end of the Permian to the end of the Jurassic (fig. 1.7). However, that figure is insufficient to indicate the increase in bioturbation that occurred during the

Mesozoic. Aside from the holothurians, individual sediment-reworking of animals with fossil records back to the Paleozoic ranges approximately from 10^{-2} to less than 10^1 cm^3 day^{-1}, whereas those that range back to the Triassic and Jurassic have individual sediment-reworking averaging from 10^1 to over 10^3 cm^3 day^{-1} (Thayer 1983). This greater intensity of disturbance by most of the more recently added bioturbating groups means that there was not just a 1:1 increase in bioturbation during the Mesozoic as new groups originated. The dashed line in figure 1.7 is a very conservative, qualitative indication of the differential increase in overall bioturbation intensity as more effective sediment-disturbing groups were added through time. It plots the sum of the order of magnitude exponent of the volume of sediment moved in cm^3 day^{-1} that characterizes each group alive at the time. The line appears as if on a semilogarithmic graph. Neither line plotted in figure 1.7 takes into account variation in population density and depth of bioturbation, nor invasion of new environments.

Occasional burrows down to depths of perhaps 1 m can be found in rocks back to the Ordovician (Sheehan and Schiefelbein 1984), but burrows of such depth are not commonly encountered until after the mid-Paleozoic. Bioturbation is widespread in shelf-depth rocks from Cambrian onward, and generally degree of disruption of sedimentary laminae increased through the Paleozoic (Droser and Bottjer 1988, 1989, 1993). Lamination in fine-grained, shelf-depth post-Paleozoic sedimentary rocks characteristically is disrupted by bioturbation except where bottom water is deficient in oxygen (Savrda and Bottjer 1991).

Morphology of burrows occupied by suspension-feeders on average is simpler and more clearly related to the sediment–water interface than that of burrows generated by deposit-feeders. In carbonates, and in silt and coarser-grained siliciclastic sediments that do not compact, the depth that burrows of suspension-feeders reached with respect to the sediment–water interface is directly preserved within rocks and often can be measured. (Assuming that one can deal with potential complications: active sedimentation during occupation and maintenance of burrows can result in longer structures than the animal occupied, and postoccupancy erosion can truncate the length.) There has been at least a three-step increase in common depth of such burrows (fig. 1.7; Ausich and Bottjer 1982; Bottjer and Ausich 1986). By Cambrian time, the earliest endobenthic suspension-feeders explored the uppermost sediment, penetrating down to about 6 cm. Then in the Late Cambrian or Ordovician suspension-feeding bivalves extended down to about 12 cm, and during the Carboniferous siphonate anomalodesmatid bivalves penetrated more deeply, reaching depths of almost 1 m (Bottjer and Ausich 1986). Most of the deep-burrowing animals still reach only about 1 m depth, but some living thalassinoid crustaceans are known to burrow to about 4 m depth (Thayer 1983).

The proliferation of bioturbators after the Paleozoic has inevitably caused a decrease in postdepositional stability of sedimentary bottoms and an increase in turbidity of near-bottom waters due to biogenically enhanced resuspension. The decreased stability and especially the increase in biogenic turbidity are hypothesized to have caused the orders-of-magnitude decrease to the virtual elimination of epibenthic sedentary suspension-feeders in most shallow sedimentary environments (Thayer 1979, 1983).

PREDATION

A second hypothesis is that increased predation has made life as an epibenthic animal more hazardous, leading to a decrease in epibenthos in general and sedentary epibenthos in particular. Predation intensity was ratcheted up from the Mesozoic to early Cenozoic. During this time, new clades and many already-existing groups of predators evolved more effective morphologies for predation than had existed earlier, causing a profound reorganization in the benthic ecosystem. This hypothesis, generally known as the "Mesozoic marine revolution," has been developed especially thoroughly by Geerat Vermeij (1977, 1987).

Data bearing on the effect of predation through time can be derived from the history of predators and from the history of prey. Data from predators include their diversity through time and the timing of apparent or inferred changes in effectiveness of predation-enabling morphology or behavior.

Diverse lineages of whole animal–ingesting and durophagous (shell-crushing) lineages were present from the Cambrian, including priapulids (Conway Morris 1977), supplemented through time by arthropods and nautiloid cephalopods (Cambro-Ordovician boundary) and in the Late Silurian by durophagous fishes (Brett 2003). New lineages of durophagous fishes—both chondrichthyes and osteichthyes—originated rapidly during the Devonian, part of the "mid-Paleozoic precursor to the Mesozoic marine revolution" (Signor and Brett 1984). Diversity of durophagous predators generally increased through the Carboniferous, declined during the Permian, then recovered by Late Triassic (fig. 1.8; Vermeij 1987). The early Mesozoic ranks of durophages were swelled by hybodontid sharks, pycnodontid and semionotiform fishes, and then during the Cretaceous by many new groups of shell-crushing malacostracan crustaceans (Harper 2003). A vigorous radiation of teleost fish during the Paleogene (Patterson 1993) added several lineages of benthic predators, including the Balistidae, a particularly effective and voracious group of tropical to temperate durophages.

Shell-chipping invertebrate predators include brachyuran crabs and various neogastropods, which use various techniques to gain access to the flesh of their prey (Vermeij 1987; Alexander and Dietl 2003). Crabs with robust

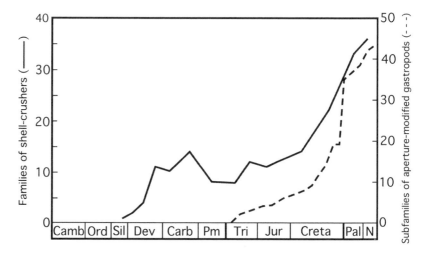

FIGURE 1.8   Indications of escalation in durophagous predation through the Phanerozoic. The solid line traces diversity of durophagous (shell-crushing) families of predators, and the dashed line traces diversity of subfamilies of gastropods with apertural modifications (lip thickening and/or narrow slitlike apertures) that inhibit peeling of the shell back from the aperture (after data in Vermeij 1987).

claws peel off portions of the shell of gastropod prey by grasping and breaking the outer margin of the snail's aperture. Some busyconine whelks attack bivalve prey such as *Mercenaria* by grasping them with their foot and repeatedly striking the commissure of the bivalve with the rim of the predator's aperture, thus producing recognizable damage on skeletons of both prey and predator (Dietl 2003a, 2003b).

Drilling or shell-boring predators also leave skeletal evidence of damage as partial or complete penetration of shell or skeleton that protects the soft tissues of the prey. Drill-holes are produced by a wide variety of predatory taxa, including diverse "worms" such as flatworms; octopods; and several lineages of gastropods (Vermeij 1987; Kowalewski 2002). The shell-bearing gastropods have left the most reliable fossil record from which geologic ranges of drilling predators can be determined. Shell-drilling gastropods are distributed among the Vayssidereidae (a dorid nudibranch family), the Muricacea (Lower Cretaceous–Recent), Hipponicidae (Lower Cretaceous–Recent), Cassidae (Upper Cretaceous–Recent), and Naticacea (Triassic–Recent) (fig. 1.9), with less extensive shell-drilling in a few other gastropods (Vermeij 1987; Kabat 1990; Kelley and Hansen 2003). The two major groups among these are the Muricacea and the Naticacea (Kabat 1990). The voracious Naticidae, which originated during the Late Cretaceous (Campanian), are the only family within the Naticacea that are predatory (Kase and Ishikawa 2003).

Other forms of predation leave no skeletal trace. Any mineralized animal that is swallowed whole but whose skeleton is not crushed (or dissolved)

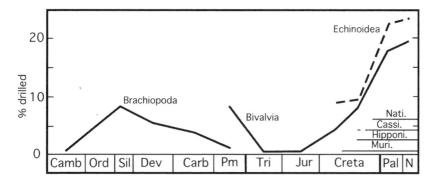

FIGURE 1.9 Mean values of percentage of drilled brachiopod, bivalve, and echinoid skeletons through the Phanerozoic (after bulk-sample data in Kowalewski et al. 1998, 1999; Kowalewski and Nebelsick 2003; Hoffmeister et al. 2004). The geologic ranges of the shell-drilling gastropods Muricacea, Hipponicidae, Cassidae, and Naticidae are given along the right edge of the figure. Identity of the Paleozoic shell-drillers is unknown. Ranges of the gastropod groups were determined from Tracey et al. 1993.

during passage through the digestive tract will emerge without any specific indication of predation. Whole animal–ingesting predators existed during the Paleozoic. These include various "worms," vertebrates that swallowed prey without any decipherable pattern of crushing (e.g., Schaumberg 1979), and possibly Paleozoic sea stars. However, a new habit of whole-animal ingestion arose within the gastropods in the Early Cretaceous and proliferated rapidly, developing within the tonnids, olivids, cymatiids, and conids (Harper 2003).

Predation by prying apart skeletal elements also need not leave any trace on the skeleton of the prey. Asteriid sea stars use their powerful sucker-bearing arms to pull apart the valves of bivalve prey. The Asteriidae originated in the Early Jurassic, but a preserved sea star spanning the commissure of an Ordovician bivalve (Blake and Guensburg 1994) raises the question of extra-oral predation by sea stars extending back to the early Paleozoic. Octopods, which have a limited and unreliable fossil record, also use suckered arms to pull apart prey. Various crustaceans and gastropods insert sharp margins between valves of prey and pry them open, essentially a post-Paleozoic phenomenon (Harper 2003).

Damaged skeletons of prey are perhaps an even better way to judge the history of shell-crushing and drilling predation than tracking the history of potential predators. Each such skeleton preserves the physical evidence of one or more predation events. In addition, changes in antipredatory morphology through time can be used to infer altered predation intensity. Predator-induced breaks in skeletons can be found occasionally in Cambrian and younger trilobites (Babcock 2003). Predator-induced breaks in more heavily mineralized shells can be traced to the Ordovician, and the incidence of such damage increased somewhat through the Paleozoic (Brett 2003; Leighton 2003). However, the evidence of attacks by durophagous predators is appreciably higher for post-Paleozoic mineralized benthos. There are numerous studies on repairs to broken shells and other mineralized invertebrates of Cretaceous through Neogene benthos that document an overall increase in skeletal repair for different groups of gastropods, bivalves, echinoderms, and other marine invertebrates (e.g., reviews in Alexander and Dietl 2003; Baumiller and Gahn 2003).

Predator-resistant changes in prey morphology have commonly been interpreted to infer elevated levels of predation. The apparent mid-Paleozoic precursor of the Mesozoic marine revolution in predation intensity marks a rapid increase in morphologies that appear to be predator resistant (Signor and Brett 1984). These include Devonian increase in spinosity of brachiopods (Alexander 1990; Leighton 2003) and crinoids (Signor and Brett 1984; Brett 2003), and the appearance of skeletal structures forming a screen over feeding structures of fenestrate bryozoans (McKinney et al. 2003), among others. Nonetheless, predator-resistant morphologies were most vigorously proliferated after the Paleozoic. One of the more striking Mesozoic through Cenozoic trends in gastropod morphology is the appearance and differential success of taxa with internally thickened or narrowed apertures (fig. 1.8), an effective defense against crustaceans that peel the apertural lip of gastropods. Umbilicate gastropods, which are more easily crushed than those lacking an umbilicus, declined slowly from Ordovician through Permian, then declined again abruptly in the Late Cretaceous (Vermeij 1987). Bivalves with radial ribs producing crenulated margins were uncommon during the Paleozoic, but the appearance of radial ribs in several bivalve lineages resulted in about 20 percent of the Jurassic through Paleocene and 35 percent of the living species having them (Vermeij 1987).

Erect bryozoans, which are more readily consumed by durophagous predators, were more diverse and common than encrusting bryozoans during the Paleozoic, but from Early Cretaceous to the present encrusting bryozoans have become by far the more prevalent in shelf-depth waters (Jackson and McKinney 1990). The cheilostomate bryozoans originated in the Late Jurassic, characterized initially by zooids with peripheral calcification and only a membrane covering the frontal surface (Taylor 1990). More extensive calcification and progressively stronger frontal shields of zooids evolved during the Late Cretaceous, and such grades of mineralization have become more diverse and abundant than those that are covered only by a cuticular membrane (McKinney et al. 2003). These are just a few from a long list of changes focused largely in the Cretaceous and Paleogene that can be inferred to indicate morphological evolution of prey in response to increase in predation.

Small, generally circular holes have been found in latest Precambrian mineralized shells and have been inferred to represent drilling predation (Bengston and Yue 1992), and predatory drill-holes are present in invertebrate skeletons of all subsequent periods of time. In Paleozoic faunas where drill-holes are found, they occur in a small percentage of brachiopod shells in the Cambrian, reach moderate percentages in the middle Paleozoic, and are low by the Permian (fig. 1.9), although high levels of muricid predation can be found in some Neogene brachiopods (Harper 2005). In the same Permian faunas where drilled brachiopods are relatively scarce, a larger proportion of bivalves are bored (Hoffmeister et al. 2004). Early in the Mesozoic, relatively few bivalve shells were bored, but beginning in the Cretaceous the proportion increased (fig. 1.9). The Naticidae are particularly active drillers (Kelley and Hansen 1993, 2003), and their presence in addition to the other shell-drilling gastropods is the apparent cause of the Late Cretaceous–Paleogene rapid increase in the occurrence of drilled shells. Echinoids are prey of cassid and other gastropods, and drilling frequency has increased from Early Cretaceous to the Recent (fig. 1.9; Kowalewski and Nebelsick 2003).

NUTRIENTS

A third hypothesis is that the more energy-intensive organisms of the Modern fauna have an advantage over those of the Paleozoic fauna in more nutrient-rich post-Paleozoic seas. Average nutrient content of marine waters and rate of nutrient cycling through marine ecosystems appear to have increased through the Phanerozoic, allowing the Modern fauna to replace the Paleozoic fauna on open marine shelves.

Thayer (1983) summarized information from the 1970s and early 1980s on nutrient recycling in the sea at present due to bioturbation. Cycling of nutrients from sediments back into the water column is enhanced by water flushing through the passageways and more loosely packed sediment generated by bioturbation, and by release of nutrients from microbial "gardens" lining some burrows. Thayer's point was that the increase in bioturbation through time, most particularly the accelerated increase during the Mesozoic, has contributed to temporal increase in nutrient levels in the seas.

The topic also was briefly discussed by Vermeij (1987) in his book centered on documenting and discussing escalation of organic interactions through geological time. At the time, he followed Thayer in attributing a long-term rise in marine nutrient levels to increased bioturbation, also noting that the invasion of land by tracheophyte plants established nutrient-rich soils from which the nutrients could be leached and washed into marginal marine environments. The Devonian invasion of relatively moist land by sphenopsid and lycopsid trees was followed by the Permian and Triassic rise of conifer forests, and then by angiosperms—some of which have nitrogen-fixing symbionts—during the Cretaceous. Vermeij later (1995) hypothesized that periods of intense submarine volcanism were the principal extrinsic cause of "revolutions" in the marine ecosystem, due to the increased input of nutrients from the volcanic activity.

Richard Bambach (1993, 1999) wrote more wide-ranging and thorough examinations of the apparent increase in availability of marine nutrients and primary productivity through the Phanerozoic. He noted seven historical patterns that collectively suggest a long-term increase in primary productivity. The first of these two papers consisted of an elaboration of data and reasoning related to the seven historical patterns, but the seven patterns can be summarized (Bambach 1993:373) as:

1. "Diversity, ecosystem complexity, and the utilization of ecospace have increased during the Phanerozoic." This essentially reviews the pattern noted in earlier work (Bambach 1983, 1985), some of which is summarized in the section on ecosystem succession above.

2. "The total biomass of marine consumers has increased." This is inferred by the post-Paleozoic increase in shell-bed thickness (Kidwell and Brenchley 1996), driven by larger numbers of benthic organisms and by increases in shell size and durability.

3. "More energetic modes of life have become common among dominant macroscopic organisms." This is a general pattern seen widely across the epibenthos throughout tropical and temperate shallow-water environments. In these environments, higher-metabolic epibenthic bivalves have replaced articulate brachiopods (Rhodes and Thompson 1993), except in unusual low-nutrient refuges (Thayer et al. 1992); cheilostomate bryozoans have replaced lower-metabolic stenolaemate bryozoans (McKinney 1993); and high force–generating predators such as crusher-clawed crabs and balistid teleosts have come into existence (Vermeij 1987; Patterson 1993).

4. "Within any particular Bauplan the effectiveness of nutrient acquisition does not increase with time," and in general increased effectiveness of resource utilization requires the development of new groups.

5. The long delay after the Cambrian in the spread of high-energy groups across relatively low-energy environments (such as the open shelves) indirectly indicates a long-term increase in nutrients, reaching levels in the Mesozoic that could support the high-energy groups in such environments.

6. "An increase in primary productivity was necessary to support the increases in biomass, metabolic rates, tiering. . . , and predation that have occurred during the Phanerozoic."

7. "Evidence of increased global productivity during the Phanerozoic comes directly from both the rise of life on land and the increase in diversity of marine phytoplankton."

It has not yet been possible to determine by direct sampling that there has been an increase in average concentration of nitrogen, phosphorus, or critical micronutrients in the seas through the Phanerozoic. The several lines of indirect evidence briefly summarized above have been elaborated by several ecologically oriented paleontologists (e.g., Vermeij 1987, 1995; Bambach 1993, 1999; Martin 1996; Allmon and Ross 2001). The weight of the evidence, each line of which is consistent with the hypothesis of marine nutrient increase, seems compelling even if any individual line may not be particularly strong.

CAN THE THREADS BE DISENTANGLED?

The characterizations of the hypotheses of cause of large-scale ecological change in the seas through the Phanerozoic, as summarized above, are not intended to be exhaustive. Instead my purpose is to give an understanding of the essence of each.

Thayer (1983), Vermeij (1987), and Bambach (1993), each of whom wrote seminal papers, have each indicated some probable degree of interrelationship among the hypotheses. Looked at as a matrix of possibilities, each hypothesis may be fully explanatory, irrelevant, or partially explanatory in combination with other(s) formulated or not yet recognized. Partisan arguments and evenhanded holistic integration of the three aside, can a judgment be made about which, if any, of the three possibilities is the most important in having shaped the evolution of the marine ecosystem?

The northern half of the Adriatic Sea is an interesting, perhaps unique, shallow inland sea with a pronounced west-to-east change in benthic life modes whose study is in many ways like peering back into time. The western benthos is solidly within the Modern fauna that typifies shelves of today's temperate and tropical oceans, whereas there are large eastern regions in which the benthos is ecologically, though not taxonomically, characteristic of the offshore Paleozoic fauna.

The purpose of this book is to characterize the oceanographic properties and benthic ecosystem of the northern Adriatic Sea, with hopes of shedding some light on the transition of the shallow benthic ecosystem from Paleozoic to Modern ecology. I begin with geography and the geological development of the Adriatic in chapters 2 and 3, then physical oceanographic properties and patterns in chapter 4, pattern of nutrient availability in chapter 5, and Pleistocene to Holocene sediment sources and distribution in chapter 6. After the oceanographic context is set in these first few chapters, composition and distributional pattern of the benthos and benthic life modes are described in chapter 7. Chapter 8 briefly examines the fidelity with which skeletal residues might represent the original ecological structure of local benthic communities. Chapter 9 first compares the change in benthic ecology across the northern Adriatic with various environmental patterns; assesses the relative contribution of depth, sediment texture, bioturbation, predation, and nutrient availability to the time-mimicking ecological gradient; and then looks at the historical transition in benthic ecology in the light of information from the northern Adriatic. Finally, the epilogue is a brief summary of current and projected changes, largely human-induced, in this remarkable shallow-water window into benthic ecology.

CHAPTER TWO

GEOGRAPHY OF THE NORTHERN ADRIATIC SEA

—Venice, lost and won,
Her thirteen hundred years of freedom done,
Sinks, like a sea-weed, into whence she rose!
—BYRON, *Childe Harold's Pilgrimage*, canto 4, stanza 13

Two narrow, shallow straits separate the Adriatic Sea from the Atlantic Ocean (fig. 2.1). The Strait of Gibraltar, which separates the Mediterranean Sea from the Atlantic Ocean, has variable width and depth along its length. It is only about 300 meters deep and a little over 20 kilometers wide north of Tangiers. Two thousand two hundred kilometers to the east of Gibraltar, the Otranto Strait separates the Adriatic Sea from the Mediterranean. It is about 800 meters deep and 72 kilometers wide. As a result, the connection of the Adriatic with the world's oceans is rather tenuous, and this, combined with its near landlocked position, might lead one to predict very different oceanographic and biological characteristics than exist along the Atlantic coasts of Europe and Africa near the Strait of Gibraltar.

REGIONS AND GEOMETRY

From its connection with the Mediterranean, the Adriatic extends north-northwest as a narrow, elongate, landlocked sea (fig. 2.2). It is bounded on the east by the Balkan Peninsula and on the west by Italy. From the Otranto Strait to Venice it is about 800 km long; south of the Gargano Peninsula it is about 200 km wide while being only slightly narrower north of the peninsula.

Three major regions from south to north are generally recognized in the Adriatic (fig. 2.2). The southern and middle Adriatic areas are divided by a

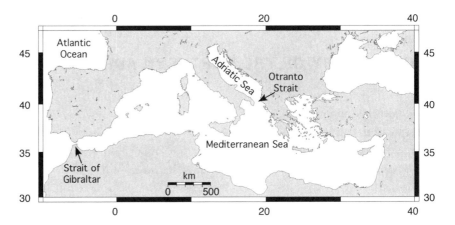

FIGURE 2.1 Map of the Mediterranean and the seas extending from it. Note the constrictions between the Atlantic Ocean and the Mediterranean at the Strait of Gibraltar and between the Mediterranean and the Adriatic at the Otranto Strait.

The base for this and other maps in this book with land areas indicated by gray shading was generated by Online Map Creation (www.aquarius.geomar.de).

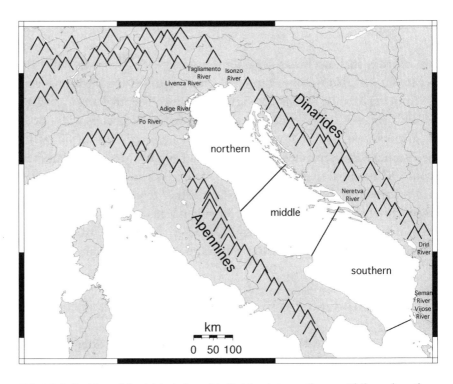

FIGURE 2.2 Map of the Adriatic Sea with division into northern, middle, and southern regions. Some important physical features of the adjoining lands are indicated.

line from the Gargano Peninsula, Italy, through Pelagosa (Croatian: Palagruža) and Sušac islands, to the Croatian coast. In contrast, the boundary between the middle and northern regions is placed more arbitrarily because there are no prominent terrestrial features to which it can be tied, nor is there such a profound difference in submarine topography as separates the southern region from the more northerly portions. Three examples of a boundary chosen to separate the two northern regions follow (see fig. 2.3 for topographical references):

1. A line from Ancona, Italy, to the southern tip of the Istrian Peninsula (Cognetti et al. 2000).
2. A line between Ancona on the west and just north of Zadar, Croatia, on the east (Buljan and Zore-Armanda 1976).
3. The essentially linear, northeast–southwest trending 100 m depth contour, which defines a line from just north of Pescara on the Italian coast to south of Zadar on the Croatian coast (Artegiani et al. 1997a).

The most logical of these three definitions is the last, in that it is based on a linear submarine trend rather than on two convenient points on land that correspond to no obvious submarine geography. However, it is not the most pragmatic for a review based largely on published literature. The boundary chosen by Artegiani et al. is farther south than the others, and few of the published studies intending to encompass the northern region extend farther south than the Ancona–Zadar line (definition 2).

The southern Adriatic contains narrow shelves surrounding a steep-walled basin that is floored by a sedimentary plain with maximum depth of approximately 1,225 m (fig. 2.3). Two submarine sills bound this basin, one across the Otranto Strait and another across the northern end, the Pelagosa Sill, which at its deepest point is about 170 m, separating the southern and middle regions (Buljan and Zore-Armanda 1976).

Appreciably shallower than the southern portion, the middle Adriatic has a central channel just over 150 m deep that extends northward from the deepest part of the Pelagosa Sill between broad, island-bearing shelves along both the eastern and western coasts, with far more islands on the eastern than on the western shelf. At the north end of the central channel is a complex transverse depression, commonly referred to by Italians as the Pomo Pit and by Croatians as the Jabuka Pit. The three primary parts of the transverse depression reach depths of between 240 and 270 m (van Straaten 1971), and the middle Adriatic averages roughly 130–150 m deep.

The northern region (figs. 2.2, 2.3) is much shallower than the more southerly regions and occupies the flooded seaward extension of the Po Plain. It slopes gently toward the south and is entirely shallow, having a maximum depth of 100 m—attained only at the boundary with the middle

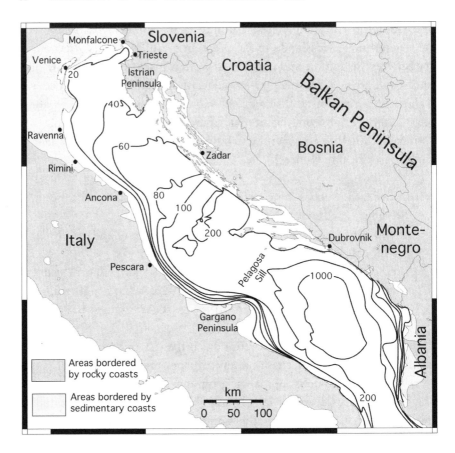

FIGURE 2.3 Bathymetric and coastal characteristics map of the Adriatic. Depth contours are in meters.

Bathymetry after Manca et al. 2002.

Adriatic—and an average depth of 35 m (Buljan and Zore-Armanda 1976; Zavatarelli et al. 2000). It is the most extensive region of shallow water in the Mediterranean and its secondary seas.

In surface area the Adriatic is approximately 139,000 km^2, and its overall depth averages 240 m, resulting in a volume of nearly 35,000 km^3 (Buljan and Zore-Armanda 1976; Zore-Armanda 1983). With an area of 57,000 km^2 and an average depth of 450 m, the southern region accounts for 28,000 km^3, about 80 percent of the total volume (Buljan and Zore-Armanda 1976). The middle region accounts for about 15 percent of the total volume, with only about 5 percent located within the northern Adriatic.

COASTS

Overall, the Balkan coast is rugged and rocky and the Italian coast has more low-lying, sedimentary stretches (fig. 2.3). Monfalcone, situated at the northernmost point of the Adriatic, marks the point of abrupt change between the sedimentary Italian coast to the southwest and the rocky coast extending southeast along the Balkan side.

From the Istrian Peninsula southward to the Albanian border, the Balkan coast marks the seaward edge of the Karst Plateau and other carbonate terrains (Celet 1973; Pavlovec et al. 1987). Aside from local gravely beaches in embayments, it consists of an essentially continuous exposure of carbonate rocks, with numerous carbonate islands offshore. The Albanian coast is largely sedimentary, consisting of alluvial plains built by several relatively small, sediment-laden rivers, two of which individually deliver more sediment at present than does the Po River (Simeoni et al. 1997). Three small headlands consisting of Neogene sediments occur between wide bays along the northern portion of the Albanian coast; the central stretch consists entirely of Holocene sediment and completes the Adriatic portion of the Albanian coast, which becomes rocky again where it faces the Ionian Sea (Simeoni et al. 1997).

A sedimentary plain borders the Italian coast from Monfalcone to Rimini, just south of which the Apennines meet the coast. The coast between Monfalcone and Rimini consists of deltas alternating with sand beaches and barrier islands, an area dominated by longshore transport before it was broken up by barriers built during the twentieth century (Colantoni et al. 1997; Simeoni and Bondesan 1997). Even the portion of the coast from Rimini to the Gargano Peninsula is bounded largely by narrow beaches where the soft sedimentary rocks of the Apennines have eroded into straight stretches between the few, well-separated, short sections where more resistant rocks form promontories and hinge points. From the Gargano Peninsula south to Otranto, the Italian Adriatic coast is almost continuously rocky.

Sources of Fresh Water

Water discharged from the Po River accounts for over 25 percent of riverine water entering the Adriatic (table 2.1), derived from the most extensive area of any of the rivers that flow into the sea (fig. 2.2). Almost 40 percent of riverine water flowing into the Adriatic enters in the short distance between the Po and the Isonzo River, the mouth of which lies a few kilometers west of Trieste, and there are no other major points of inflow of fresh water in the entire middle and northern regions of the Adriatic (fig. 2.4). The edge of the Karst Plateau and other carbonate masses are exposed along the Istrian and Dalmatian coasts between Trieste and Dubrovnik, and much of the seaward flow of fresh water in this region is subterranean, through the porous carbonate rock (Poulain and Raicich 2001).

TABLE 2.1 Primary Rivers Entering the Adriatic and Their Mean Annual Discharge Rates

RIVER	DISCHARGE IN m³ sec⁻¹
Po	1,585
Adige	220
Livenza	88
Tagliamento	97
Isonzo	204
Neretva	378
Drin	338
Seman	200
Vijose	182
All other surface drainage	2,384
Total	5,676

See fig. 2.2 for locations of the primary rivers.
Data from Raicich 1994.

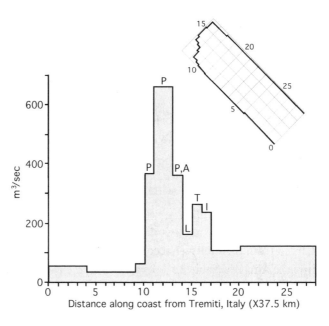

FIGURE 2.4 Ten-year average surficial inflow of fresh water into the Adriatic per unit length of coast for November through December in the middle and northern Adriatic, from Tremiti Archipelago, Italy, to the island of Korčula, Croatia. (A line from Tremiti Archipelago to Korčula follows the Pelagosa Sill.) Rivers emptying into coastal segments 11 through 17 are the Po (P), Adige (A), Livenza (L), Tagliamento (T), and Isonzo (I) (see fig. 2.2 for locations of rivers).

Modified from Cavazzoni Galaverni 1972.

Winter is driest and autumn wettest in the northernmost lands bordering the Adriatic, but summer is driest and winter the wettest along the rest of the Adriatic borderlands (Raicich 1996). Although there are differences in when wet and dry seasons occur in the northern and southern drainage basins, there is a basinwide annual pattern of minimum runoff during summer and essentially uniform runoff from autumn through spring (Raicich 1996).

Because it gains more fresh water than the mass of water lost by evaporation, the Adriatic is a net exporter of water into the Mediterranean. The freshwater balance depends upon the total input, which includes both direct precipitation onto the sea as well as that gained by flow into the sea from the adjacent land, and the total loss, which occurs only by evaporation. Based on a large set of data and on modeling of evaporation pressure, Raicich (1996) estimated average annual evaporation rate across the Adriatic to be 1.08–1.34 m, average direct precipitation to be 0.82–1.02 m per year, and annual runoff equivalent to be 0.91–1.42 m. He therefore inferred an annual excess freshwater accumulation of 0.65–1.10 m. For the entire basin, this equals about 90–150 km^3 net gain of fresh water per year that is exported through the Otranto Strait into the Mediterranean.

DISPLACEMENT OF THE SHORELINE

A single, partially flooded sedimentary plain constitutes the floor of the Po Plain and the northern Adriatic. The broad valley occupied by the Po and adjacent rivers declines gently toward the east at a slope of about 1:2,000 before emptying into the northwest corner of the Adriatic Sea, which slopes gently toward the southeast at less than 1:10,000 (Brambati 1992). This gentle, continuous slope from the Po Plain into the sea permits large lateral movement of the coastline with the slightest change of sea level relative to level of the land surface.

Glacially driven eustatic changes in sea level during the Pleistocene and Holocene, local tectonically driven changes in elevation, and coastal sedimentation all have played roles in causing significant changes in the extent of the northern Adriatic Sea during the Pleistocene and at present. Only the most recent changes (Holocene) are summarized here. Longer-term geological development of the Adriatic basin is discussed in chapters 3 and 6.

POTENTIAL GLACIALLY INDUCED SEA-LEVEL CHANGES

Maximum Late Pleistocene glaciation resulted in worldwide sea level about 130 m lower than at present (Lambeck et al. 2002), and complete melting of glaciers would raise sea level about 50 m higher than at present. In the

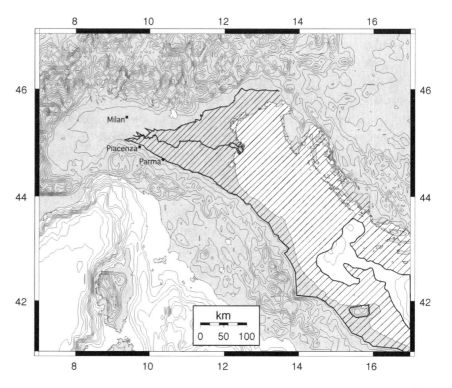

FIGURE 2.5 Minimum and potential maximum extent of the northern and middle Adriatic due to glacially driven changes in sea level. The area encompassed between maximum and minimum is approximated by diagonal hatching. Shoreline for minimum extent of the sea is based on the extent of the Adriatic Sea during the last glacial maximum as portrayed in Asioli et al. 1996. Potential maximum extent traces the present 50 m contour.

absence of any other influences, change in sea level due to maximum glaciation would cause a draining of the entire present northern Adriatic Sea, below the 100 m contour that marks the boundary between the northern and middle regions of the sea (fig. 2.5). Alternatively, complete melting of the existing glaciers would flood what is now the Po Plain almost to the longitude of Milan (fig. 2.5).

These are more than theoretical possibilities. Early Pleistocene marine deposits of the Adriatic Sea can be found along the northern edge of the Apennines, extending west of Parma and up to elevations of 185 m (Di Dio et al. 1997; Taviani et al. 1998). The high local elevation of these deposits is due to active tectonic deformation of the Apennines, but their occurrence so far to the west is the point here. They indicate that Pleistocene eustatic changes in sea level have indeed caused the Adriatic to spread far west of the present shoreline during interglacial intervals.

During the last Pleistocene glacial interval (the Würm), which lasted until about 18,000 years ago, the entire northern Adriatic region was a fluvio-

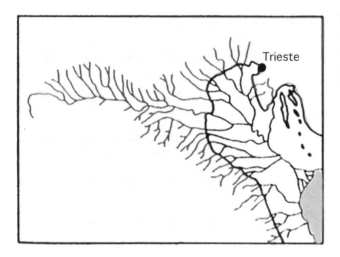

FIGURE 2.6 Paleogeographic sketch of the river system that extended across the current northern and much of the middle Adriatic Sea during the Würm glacial interval.

From Ferretti et al. 1986.

lacustrine plain (fig. 2.6; Colantoni et al. 1979; Ferretti et al. 1986). Such profound emptying of the Adriatic is also indicated by an erosion surface off the Gargano Peninsula that is presently about 112 m deep, and estimated to have formed only about 18,000 years ago at the close of the Würm glaciation (Pasini et al. 1993). Sea-level rise since then generated a transgression of the shoreline that has left a series of coastal erosional and depositional features across the northern region, especially a well-developed former shoreline about 25 m deep (Colantoni et al. 1979). The encroaching Holocene shoreline apparently reached its maximum extent thus far about 5,500 years ago, as indicated by a well-dated relict shoreline that lies north and west of the lagoon of Venice and the Po Delta (Correggiari et al. 1996b, 2001; Stefani and Vincenzi 2005).

The Holocene record is most voluminous in and offshore of the Po Delta, but it is also well developed in more northerly locations such as the lagoon of Venice. Several boreholes in the lagoon, along a marshy bank between Torcello and Canale San Felice, in the vicinity of Burano, penetrate Holocene sediments that vary from 1.5 to 6.0 m thick (Favero and Barbero 1981). These sediments overlie Würmian fluvial sands and clays and consist of several stacked facies: fine-grained sediments with abundant fossils that indicate an outer lagoon environment; a zone of increasing grain size and decreasing diversity of macrofauna, with low-salinity gastropods near the top; a coarser-grained interval with relatively open marine foraminiferans; and finally a zone of alternately fine and relatively coarse sediments with at first scattered and at the top ubiquitous plant remains indicative of salt marsh like that currently present on the bank (Favero and Barbero 1981). This sequence is interpreted as transgressive up to the point of the coarse-grained interval with relatively open marine foraminiferans, and regressive from that level to the present-day conditions.

Littoral sand just below the point of maximum transgression contains Roman artifacts and can be dated by correlation with similar deposits elsewhere to the third and fourth centuries A.D. Within this local section, the point of maximum transgression, that is with the shoreline farthest inland, probably occurred in the fifth century, followed by regression and eventually, as indicated by numerous historical documents, with the formation of the salt marsh after 1500 A.D. (Favero and Barbero 1981). Such shifting of shorelines and evolving environments during the Holocene suggests that the margin of the northern Adriatic is presently in a state of dynamic change.

CURRENT SEA-LEVEL CHANGES

At present, sea level and position of the shoreline are changing in the northern Adriatic due to several worldwide, regional, and local causes. A worldwide (eustatic) sea-level rise of several centimeters has apparently occurred over the past 100+ years (e.g., Shennan and Woodworth 1992). Satellite altimetry—supported by coastal tide gauges—documented an impressively rapid sea-level rise of between ~10 mm and ~20 mm per year across the Mediterranean Sea from 1993 through 1999, apparently related to thermal expansion of deep water masses (Cazenave et al. 2001). This rise was not uniform across all of the Mediterranean and its satellite basins; in the Adriatic it varied from 5 mm per year at the Otranto Strait to >20 mm per year in the north.

Venice, a World Heritage Site, is built on a series of low islands with a mean elevation of 1 m above sea level, separated by canals. Consequently any change in the short-term or long-term level of the sea is of intense interest here. Low-lying portions of the city are subject to flooding during periods called "acqua alta" (high water), when the Piazza San Marco and the campos as well as the streets become wave-paved. Acqua alta events (fig. 2.7) became progressively more common toward the end of the twentieth century (Pirazzoli and Tomasin 2002).

Individual acqua alta events are the result of the interaction of tides, wind stress, and gradients in atmospheric pressure along the Adriatic and across the western Mediterranean that generate a bulge of water along the northern shore of the Adriatic (e.g., Bargagli et al. 2002). However, the sea-level context in which they occur determines their overall frequency. Through the twentieth century there was an overall quarter-meter loss in land elevation in Venice (Gatto and Carbognin 1981; Carbognin et al. 2004, 2005).

Some of the subsidence of recently deposited fine-grained sediment, such as that in the lagoon of Venice, is natural. Natural subsidence occurs in Venice and all of the recently sediment-accumulating coast between the

FIGURE 2.7   Acqua alta in Venice, viewed from Ponte della Paglia. The paved pedestrian area between the line of gondolas (*left*) and the Doge's Palace (*right*) is flooded by an influx of water from the Adriatic Sea; during such acqua alta events, people are confined to walking on temporary elevated boardwalks.

Photograph courtesy of Sarah Quill.

Isonzo River near Trieste and the coast south of the Po Delta (fig. 2.8). Compaction and de-watering of muds because of the weight of the overlying sediment, enhanced by electrochemical compaction due to salinity changes, are partially responsible for the natural subsidence. But as discussed in chapter 3, the entire area between the Apennines and the Alps-Dinaric Alps is tectonically unstable, with a complex distribution of currently active vertical land movement (fig. 2.9) on an overall southwesterly-steepening monocline that may account for a loss of 1 mm yr^{-1} in Venice (Carminate and Doglioni 2003). The total natural temporal variation in elevation, combined with eustatic sea-level rise, resulted in about a 12 cm loss of relative land elevation in Venice through the twentieth century (Gatto and Carbognin 1981).

Human-generated subsidence during the twentieth century was highly variable in the different areas of Venice but on average approximately equaled natural loss of elevation relative to sea level (fig. 2.10). Subsidence in Venice and nearby localities due to water extraction from artesian wells reached alarming rates of up to 10 mm per year in the third quarter of the twentieth century, before action was taken to ameliorate the cause (Gatto and Carbognin 1981; Brambati 1992). There had been over 400 artesian

FIGURE 2.8 Natural land subsidence during the late twentieth century for the northwestern Adriatic region.

From Gambolati et al. 1998, figure 1.8, with kind permission of Springer Science and Business Media.

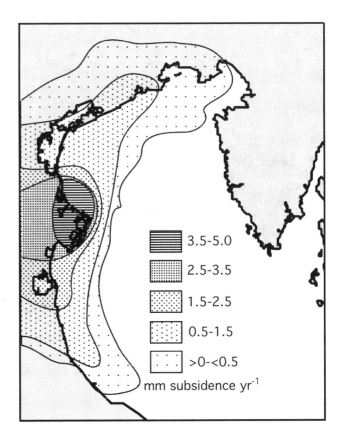

FIGURE 2.9 Vertical land movement (in mm yr⁻¹) across the Po Plain and adjacent areas from 1897 to 1957.

Based on *Carta altimetrica e dei movimenti verticali del suolo della pianura Padana*, S.El.Ca., Florence, Italy, 1997.

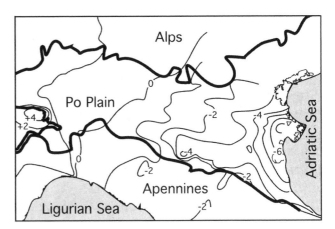

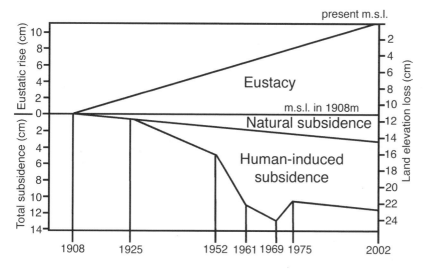

FIGURE 2.10 Land elevation loss and its apparent causes in Venice during first three quarters of the twentieth century.

Modified from Gatto and Carbognin 1981 and Carbognin et al. 2005.

wells in the municipality of Venice, but the number was reduced between 1970 and 1975 to only 22, which collectively were extracting only a trivial amount of water (Brambati 1992). This decrease in subterranean water extraction produced a coeval elevational rebound of 2 cm (fig. 2.10), and subsequent change in sea level at Venice has been more essentially similar to that in the Gulf of Trieste, which can serve as a nearby standard (Pirazzoli and Tomasin 2002).

SEDIMENT SUPPLY

For sedimentary shorelines, as exist along the northwestern Adriatic coast, especially where the elevation gradient is low, landward or seaward migration of the shore is a function of the interaction of change in sea level and sediment supply or removal.

Italian rivers flush 4.2×10^7 tons of sediment into the Adriatic each year, and among these the Po is far and away the main source (Buljan and Zore-Armanda 1976; Trincardi et al. 1994). About two thousand years ago the sedimentary coast of the northwestern Adriatic began prograding, built seaward largely by sediment transported down the Po River, which has extended into a protruding delta characteristic of coasts where sediment delivery is faster than longshore or basinward removal. From the seventeenth into the twentieth centuries, the Po Delta was prograding at 70 m yr^{-1}, over

TABLE 2.2 Decrease in Bed–Load Sand Supplied from Four Italian Rivers Into the Adriatic Sea

RIVER	BED LOAD IN M^3YR^{-1}		YEARS
Po	from	1.42×10^6	1956–1964
	to	0.87×10^6	1965–1973
Adige	from	29×10^3	1922–1950
	to	22×10^3	1958–1975
Metauro	from	51×10^3	before 1966
	to	$25–36 \times 10^3$	early 1980s
Tronto	from	74×10^3	before 1966
	to	41×10^3	early 1980s

From Brambati 1992.

an order of magnitude faster than it had been before the deforestation of the Po Plain and the construction of levees, which began about 1300 and culminated about 1600 A.D. (Colantoni et al. 1979; Oldfield et al. 2003). Construction of hydroelectric dams (which trap sediment) and dramatic increase in industrial dredging of sand from riverbeds during the twentieth century has reduced the sediment input of Italian rivers into the Adriatic (table 2.2). This reduction in sediment supply occurred during the latter half of the twentieth century and caused a switch from prograding to retreating shorelines (Colantoni et al. 1997; Simeoni and Bondesan 1997).

SUMMARY

The physical geography of the northern region of the Adriatic sets the stage for oceanographic conditions and a biota that are appreciably different and potentially much more variable than the norm for shallow-water marine conditions in temperate latitudes. First, exchange of water with the world oceans is restricted by the Strait of Gibraltar at the entrance to the Mediterranean Sea and then by the Otranto Strait at the entrance to the Adriatic. Within the Adriatic, the northern region is the most land-locked and therefore potentially is more influenced by the effects of terrestrial weather. It also is the shallowest region and therefore contains the smallest volume of water, further restricting circulation and potentially compounding the effects of perturbations. Finally, it receives a large volume of freshwater input, which can cause large variations in salinity, temperature, and nutrient levels when circulation becomes ineffective.

Extensive landward or seaward shift of the shoreline occurs where there is a low topographic gradient, such as the one continuous throughout the length of the Po Plain and the northern Adriatic, that interacts with eustatic changes in sea level and with local phenomena such as vertical tectonic movements and sediment compaction. In fact, the melting of the most recent Pleistocene ice sheets and consequent sea-level rise during the Holocene are the proximate reason that the Adriatic Sea spread across the region now known as the northern Adriatic, as well as most of the middle Adriatic. Sediment compaction and vertical tectonic movements have had a real but smaller effect in these regions recently.

The rate of sediment supply to the northern Adriatic coast, largely from the Po River but also through several smaller Italian rivers, interacts with the dynamic level of the sea to influence whether the coastline is retreating or advancing. Human activities within the rivers' drainage basins have dramatically affected the sediment supply during the past several hundred years, resulting in changing rate and even reversal of coastline movement. This is but one of the anthropogenic effects on the northern Adriatic, and several of the others that impinge on the ecosystem will be taken up in remaining chapters.

CHAPTER THREE

ORIGIN OF THE ADRIATIC

The Mediterranean is a festival of microplates.
—CHARLOTTE SCHREIBER, who heard it at a conference

Active mountain belts almost entirely surround the Adriatic Sea and its northern alluvial plain extension. The Apennines lie to the southwest, the Alps to the north and northwest, and the Dinarides along the northeast. It is well separated from mountains only along the southern third of the Italian coast, where an oblique, southeastwardly projecting peninsula (Puglia) intervenes. The distribution and geology of the enfolding mountain belts, along with the geology of the underlying seafloor, are key elements for understanding the origin of the Adriatic basin.

THE MEDITERRANEAN: STRAINED RELATIONS BETWEEN AFRICA AND EURASIA

As the Mesozoic era began, the Earth's continental masses had assembled into one huge continent, Pangea, with present-day Africa and Eurasia in contact along what would eventually become Algeria and the Iberian Peninsula (fig. 3.1). East of this point of contact the Tethys Sea occupied a large embayment into the supercontinent. The northern shore of the Tethys extended along what would become Eurasia, and its southern shore was formed by land now known as Africa, Australia, Antarctica, and New Zealand. Although the western tip of the Tethys Sea extended to what is

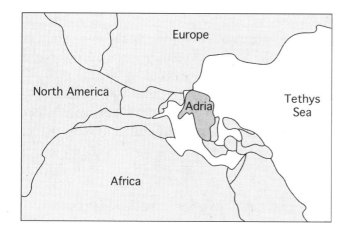

FIGURE 3.1 A reconstruction of the Sinemurian (Lower Jurassic) Tethyan realm, simplified from Schettino and Scotese 2002. Outlined gray areas other than Africa, North America, and Europe are continental parts of microplates.

now Algeria and Iberia, the Mediterranean Sea was eventually defined by differential motion of Africa and Eurasia relative to one another and by fragmentation and redistribution of small parts of the northern edge of Africa. The Mediterranean is therefore not solely a remnant of the Tethys Sea (Smith 1971; Dewey et al. 1973).

When Africa began moving relative to Eurasia, about 160–165 million years ago (Bill et al. 2001), it followed a counterclockwise path that eventually brought it to its present position. Over time the northeastern edge of Africa apparently broke into 15 to 20 small blocks, or microplates, that were caught in the differential motion of Africa and Eurasia. Several of the microplates eventually moved north and were accreted onto the southern border of Europe. One of the microplates is variously known as Apulia, the Adriatic plate, or Adria (fig. 3.1), the latter being a term used by Suess and Canavari in the late nineteenth century for a hypothetical emergent area that existed previously where the Adriatic Sea exists today. Channell et al. (1979) resurrected the term as a name for the area encompassing the Adriatic Sea and the surrounding orogenic belts, i.e., the Apennines, southern Alps, Dinarides, and Hellenides (fig. 3.2 below). (N.B.: The concept of Apulia includes a substantially larger area than does the Adriatic plate, e.g., Jolivet and Faccenna 2000.)

There is still vigorous disagreement about whether Adria became separated from the African plate (e.g., Dewey et al. 1973; Anderson and Jackson 1987; Robertson and Grasso 1995; Jolivet and Faccenna 2000; Battaglia 2004; Grenerczy et al. 2005) or is a northern promontory that remains a part of Africa (e.g., Channell et al. 1979; Mele 2001; Oldow et al. 2002; Rosenbaum et al. 2004). The presence of ocean crust between Adria and Africa, as well as the African plate's northward subduction at the Ionian Arc, indicates that Adria must have detached from Africa as an independent microplate at a relatively early time and that the gap may now be closing (A. Smith, personal communication) or has remained as an essentially unchanged link between the two since perhaps the Jurassic (Rosenbaum et al. 2004). Some argue that detachment of Adria from Africa is indicated by an approximately 30° counterclockwise rotation of Adria relative to Africa and stable Europe since the Oligocene (e.g., Márton et al. 2003), while others argue that no such rotation has occurred (e.g., Babbucci et al. 2004; Rosenbaum et al. 2004).

REGIONAL STRUCTURAL CONTEXT OF THE ADRIATIC

Extraordinary complexity characterizes the geology and geologic history of the southern border of Europe and of the Mediterranean in general. This complexity derives largely from the complicated differential movements of the microplates in the region and from geometric irregularities of the borders of colliding continental crust. There have been multiple periods of faulting due to changing conditions of compression, tension, and shear, and the resulting faults are variously oriented. Although many of the details of the history of Adria are not clear at present, enough is known to give a general idea of the origin and development of the Adriatic basin.

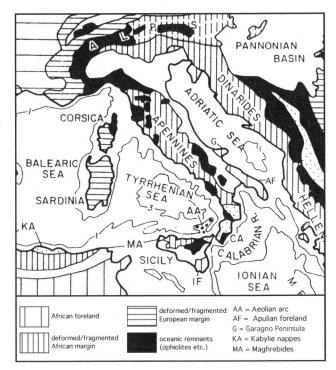

FIGURE 3.2 Large-scale tectonic features of the Adriatic margins and nearby areas. Note that African foreland rocks are exposed on the Apulian foreland and Gargano peninsula (they also constitute the basement of the Adriatic basin), that the Adriatic is almost entirely surrounded by deformed/fragmented rocks of the African margin and associated oceanic remnants, and that Sardinia and Corsica are comprised of deformed/fragmented rocks of the European margin. Compare the distribution of tectonic features in this figure with the generalized diagram of tectonic zones in figure 3.3.

From Malinverno and Ryan 1986. Reproduced by permission of American Geophysical Union.

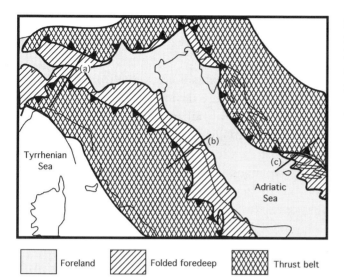

FIGURE 3.3 Major tectonic zones influencing the Adriatic Sea. Straight lines across tectonic boundaries represent paths of cross sections in fig. 3.4.

After Brambati 1992.

A general tectonic history of Adria can be gotten from the broad-scale structural elements seen in figures 3.2 and 3.3. Local oceanic remnants to the northeast and northwest of Adria record the destruction of ocean floor that previously separated Adria from Europe. The circum-Adriatic mountain belts reflect the compression and fracturing of the margins of Adria where it collided with Europe to the northeast, north, and northwest. Now it is being compressed from the southwest by growth of the Tyrrhenian plate, which lies west of Italy. By virtue of thrusting on all sides toward Adria, the Adriatic basin is now floored by a small remnant of Adria that is depressed by active loading along its margins.

Collision of Adria and Europe

Development of the northern Apennines and southern Alps began in the Late Oligocene. During the Jurassic to Early Paleogene period, the ocean basin that had separated Adria and Europe gradually closed as the Tethyan oceanic crust was subducted below the northern and eastern edges of Adria, leading to the Middle Eocene collision of Adria and Europe (Ricci Lucchi 1986; Boccaletti et al. 1990; Cibin et al. 2001; Finetti et al. 2001), reflected in the Dinaride mountains. Adria dips gently northeastward under the overriding Dinarides (Prelogović et al. 1995). Thrust faulting of the northern Dinarides over the margin of Adria apparently ceased or significantly diminished during the Oligocene epoch (Aljinović et al. 1984;

Kruse and Royden 1994), although the southern Dinarides–Adria border southward from Split remains more active (e.g., Udias 1985; Oldow et al. 2002).

A still-active, complex crush zone extends across the Dolomite Mountains in Italy and through the Northern Calcareous Alps of Austria, where the northern edge of Adria has overridden the Eurasian plate (e.g., Gebrande et al. 2002). Both the southern Alps and the northern Apennines have stacked thrust sheets that are progressively younger toward the converging boundaries of the two mountain chains and that are partially covered by sediments of the Po Plain (fig. 3.4, section A). Maximum compression between Adria and the Alps began in the northwest and progressed southeastward and eastward from Middle Oligocene time to the present, as indicated by sequential eastward shifting of maximum depth of depocenters, southeastward imbrication of thrust faults, and by right-lateral movement along strike-slip faults that cut the southern edge of the Italian Alps (Massari et al. 1986).

Northwestern Adria has therefore been progressively reduced in width and thickened on both the north and south sides of the Po Plain by the imbricated, convergent thrusts of two mountain chains, the Alps and the Apennines. Each thrust in turn caused downbuckling of the overridden region, generating a foredeep basin into which sediments were washed. Foredeeps on the Adria side of both the southern Alps and the Apennines were on average deeper during Late Oligocene and Early Miocene time than during Middle and Late Miocene time, and turbidites and other flysch sediments, derived largely from the Alps, accumulated in them. Although there was a transitional period from Early through Middle Miocene time with both shallow and sometimes distally deep basins (Rizzini and Dondi 1980; Cibin et al. 2001), Middle Miocene through Pliocene foredeeps were generally filled by molasse sequences of shallow marine and terrestrial sediments, derived predominantly from the Apennines (Ricci Lucchi 1986; Boccaletti et al. 1990).

The Original Risorgimento: Unification of Italy by Opening of the Tyrrhenian Sea

Cross sections through the central Apennines and into the northern Adriatic Sea show a generally similar pattern to that of the Apennines farther north (fig. 3.4, section B) but features are on average younger than in the more northerly Apennines. Here, the foredeep deposits of flysch in the older thrust sheets are Middle–Late Miocene in age, and the foreland deposits of molasse on the younger (more eastern) thrust sheets are Late Miocene to Quaternary (Boccaletti et al. 1990). Eastward migration of the

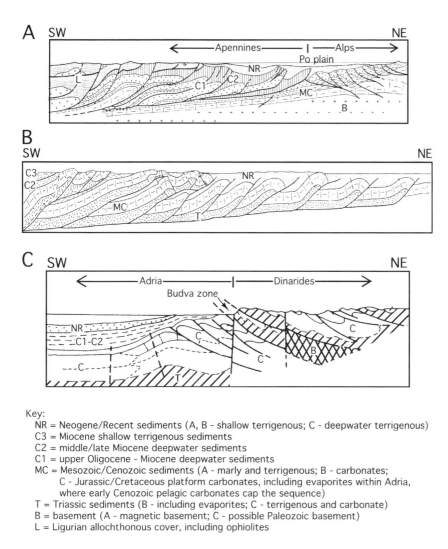

Key:
    NR = Neogene/Recent sediments (A, B - shallow terrigenous; C - deepwater terrigenous)
    C3 = Miocene shallow terrigenous sediments
    C2 = middle/late Miocene deepwater sediments
    C1 = upper Oligocene - Miocene deepwater sediments
    MC = Mesozoic/Cenozoic sediments (A - marly and terrigenous; B - carbonates;
         C - Jurassic/Cretaceous platform carbonates, including evaporites within Adria,
         where early Cenozoic pelagic carbonates cap the sequence)
    T = Triassic sediments (B - including evaporites; C - terrigenous and carbonate)
    B = basement (A - magnetic basement; C - possible Paleozoic basement)
    L = Ligurian allochthonous cover, including ophiolites

FIGURE 3.4 Structural sections showing large-scale structures in tectonic zones influencing the Adriatic Sea. See fig. 3.3 for paths of the sections. Structural section A covers a length of about 175 km and crosses the convergent boundary (buried beneath the Po Plain) between the Alps and the northern Apennines. Structural section B covers a length of about 125 km and crosses from the tectonic core of the Apennines into the relatively undeformed center of the northern Adriatic Sea. Structural section C covers a length of about 125 km and crosses from within the Dinarides into the relatively undeformed center of the southern Adriatic basin (see fig. 3.9).

Sections A and B reprinted from *Palaeogeography, Palaeoclimatology, Palaeoecology*, vol. 77, Boccaletti et al., "Migrating foredeep–thrust belt system in the northern Apennines and southern Alps," pp. 3–14. Copyright © 1990, with permission from Elsevier. Section C from Celet 1973.

thrust belt closely followed eastern migration of the axis of the adjacent, asymmetrical foredeep (steeper on the west side), which apparently was the result of a downbuckled wave due to tectonic loading on the west (e.g., Crescenti et al. 1980). Restored cross sections, in which fault blocks are conceptually moved back to a point at which the various broken sheets of rock match as continuous sheets, indicate only about 20 percent shortening of the region from the easternmost Apennines into the western portion of the northern Adriatic Sea (Coward et al. 1999), although total shortening across the Apennines may be about 30 percent (~170 km; Roure et al. 1991; Finetti et al. 2001).

The Tethyan oceanic crust was subducted under Adria's northern and western margin until mid-Eocene time (Finetti et al. 2001). Then a small part of the Tethyan oceanic crust, now known as the Ligurian sheet, was overthrust onto the northwestern margin of the thrust stack bordering Adria during Late Eocene time (Ricci Lucchi 1986; Finetti et al. 2001; see fig. 3.5A). This was followed by reversal of the plane of subduction on the west side of Adria, resulting in ancient seafloor and eventually the margin of Adria being subducted westward, i.e., below the plate that lies to the west of Adria.

Change from eastward to westward subduction along the western margin of Adria was approximately coincident with the Early Oligocene development of a spreading center that initiated the Balearic basin and separated Corsica and Sardinia from their previous positions adjacent to France (fig. 3.5: compare parts A and B). Opening of the Balearic basin coincided with the first pulse of deformation of the northern Apennines at 35 Ma (Ma = mega-annum = 10^6 yr) (Malinverno and Ryan 1986). The Balearic basin continued to expand in width until about 9 Ma (Late Miocene, Tortonian), at which time major spreading began in the Tyrrhenian basin (Malinverno and Ryan 1986; Dewey et al. 1989; Kastens and Mascle 1990; Robertson and Grasso 1995).

The westward-descending Adria plate apparently caused back-arc spreading to develop in the floor of the Tyrrhenian Sea in Late Miocene time (figs. 3.5C, 3.6), which enhanced the eastward component of compression of the western margin of Adria. The widening of the Tyrrhenian Sea (fig. 3.5C, D), associated with apparent counterclockwise rotation of Adria, contributed to the Miocene to present rise of the Apennines, progressively eastward migration of thrust-fault development, and rotation of large blocks caught in the northern hinge (Patacca et al. 1990; Roure et al. 1991; Finetti et al. 2001; Márton et al. 2002a, b). On the western side of the eastwardly migrating zone of thrusting, downflexure of the descending edge of Adria results in normal faulting that cuts across the thrust sheets. The migrating zone of normal faulting incorporates the core of the Apennines and has reached almost as far east as the middle

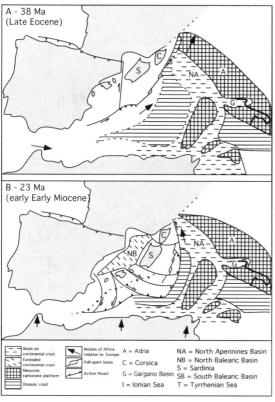

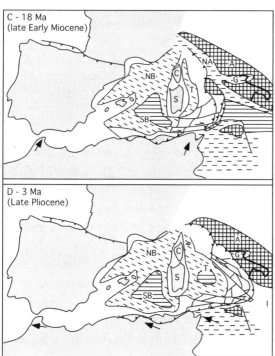

A - 38 Ma
(Late Eocene)

B - 23 Ma
(early Early Miocene)

Legend:
- Basin on continental crust
- Extended continental crust
- Mesozoic carbonate platform
- Oceanic crust
- Motion of Africa relative to Europe
- Pull-apart basin
- Active thrust

A = Adria
C = Corsica
G = Gargano Basin
I = Ionian Sea

NA = North Apennines Basin
NB = North Balearic Basin
S = Sardinia
SB = South Balearic Basin
T = Tyrrhenian Sea

C - 18 Ma
(late Early Miocene)

D - 3 Ma
(Late Pliocene)

FIGURE 3.5 Opening of the Balearic and Tyrrhenian basins, and subduction on the west side of Adria. Note that during the Late Eocene (A), Corsica, Sardinia, and other western Mediterranean islands were contiguous with the European mainland, and that by the beginning of the Miocene (B) a spreading center that developed into the North Balearic Basin began to separate the islands from Europe. During the Early Miocene the North Balearic Basin widened to essentially its present configuration (compare section C with B and with D), and a new spreading center initiated the basin in which the Tyrrhenian Sea is located, east of Corsica and Sardinia. The disparate elements that constitute Italy south and southwest of the Po Plain had assembled into a form generally recognizable as today's geography during the rapid growth of the Tyrrhenian Basin during Miocene and Pliocene (compare sections C and D).

After Dewey et al. 1989.

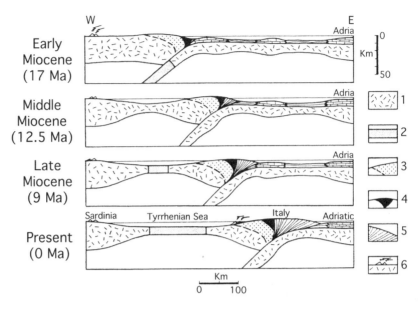

FIGURE 3.6   Possible development of the back-arc spreading center in the Tyrrhenian Sea and development of the Apennines over the subduction hinge of Adria.

From Malinverno and Ryan 1986. Reproduced by permission of American Geophysical Union.

Adriatic coast (Vezzani and Ghisetti 1998; Ghisetti and Vezzani 1999, 2002a, b; Galadini and Messina 2004).

PRESENT TECTONIC CHANGES IN AND AROUND THE ADRIATIC

Several lines of evidence can be used to interpret current stress fields and crustal deformation. These include analysis of earthquake distributions, orientation of seismic wave propagation, borehole breakouts, measured changes in distance across structures, and cross-cutting relations. While there isn't complete agreement on the meaning of different data sets, an overall pattern seems to emerge.

Analysis of first motion fault plane solutions for more than fifty large earthquakes around the Adriatic basin yields a clear northeast–southwest pattern of slip in both the Apennines and the Dinarides, but north–south slip in the eastern Alps (Anderson and Jackson 1987). The pattern observed by Anderson and Jackson (1987) suggested current counterclockwise rotation of Adria relative to Eurasia, with extension (normal faulting) at a maximum within the core of the southern Apennines approximately equal to crustal shortening (compressive movement on thrust faults) on the Balkan coast of the central and southern Adriatic Sea. Anderson and Jackson (1987) estimated about 2 mm yr⁻¹ of movement toward the northeast for the part of Adria south of the Tremiti shear zone, which is a weak northwest–southeast lineament of seismicity crossing the Adriatic from the Gargano Peninsula, and they as well as Grenerczy et al. (2005) tentatively suggested internal deformation of Adria along that line.

Deep borehole deformation in anisotropic stress fields yields elongation, known as breakouts, in the direction of minimum horizontal stress. Information from breakouts can be resolved in conjunction with orientation of local fault planes to determine if faults are essentially extensional, compressional, or transpressional (laterally moving). Analysis of breakouts and earthquakes across the Apennines and into the Adriatic basin indicates that the deep, nearly horizontal faults extending from below the Apennines into the Adriatic basin are compressional north of 43° and extensional along currently active faults behind the frontal zone (Montone et al. 2004). The compressional regime extending under the Adriatic Sea suggests to Mariucci et al. (1999) that the Apennine margin is still actively moving eastward, and the normal faulting west of this zone is consistent with tension due to stretching of the crust over the downward flexure of the subducting Adriatic plate.

Some evidence, however, suggests that the northern part of Adria is not presently being subducted, although the southern part continues to be. In the northern and central Apennines, the most distal thrust faults are draped by unbroken Late Pleistocene to Holocene sediments, and there has been only moderate seismicity on inland thrusts (Di Bucci and Mazzoli 2002). There has been no apparent compressive northeast–southwest movement on these outlying northern Apennine thrusts during Quaternary and Holocene time (Di Bucci et al. 2003). The Martana Fault within the interior of the northern Apennines has several features indicative of northeast–southwest compression up through at least 0.4 Ma, but it changed to sinistral strike-slip postdating 0.26 Ma and continues to cut through and offset historical structures, indicating that for the last 250 Ka (Ka = kilo-annum = 10⁶ yr) the northern part of Adria has been sliding northwest relative to the Italian promontory of the Eurasian plate (Bonini et al. 2003).

In contrast, several lines of evidence from southern Italy, including GPS location data, indicate continued northeast-convergence of Adria on Europe at about 2 mm yr⁻¹ or a bit more (Di Bucci and Mazzoli 2002; Grenerczy et al. 2005), as hypothesized by Anderson and Jackson (1987). The present apparent lack of compression on Adria-verging thrusts along the middle and northern Adriatic margin and the movement of southern Italy toward the Balkans support the seismic pattern that suggests that Adria has broken essentially along the Tremiti shear zone (Oldow et al. 2002). Earthquakes are much more frequent along the Tremiti shear zone and along the eastern and western boundaries of Adria south of the shear zone than they are along Adria's northern, western, and eastern borders north of the shear zone.

FOREDEEP DOWNBUCKLING OF ADRIA: THE ADRIATIC MARINE BASIN

Adria was flexed down to the east and northeast as it was subducted below the overriding Dinarides and Hellenides; it was also flexed down to the southwest, where it is being subducted below the Apennines (fig. 3.7). The southern Alps reflect at least some continued compression along the northern edge of Adria, with Alpine-rooted thrust faults cutting Neogene sediments accumulated in the Venetian basin (Massari et al. 1986).

The Venetian basin is at the northern end of the Adriatic, partially flooded by the sea and partially occupied by the Venetian plain, at the flexure between the northwest–southeast extent of the Adriatic and the east–west trend of the Po and associated valleys. The Venetian basin began as the northernmost, former foredeep basin of the Dinarides, although more recently it has been a foredeep basin of the southern Alps (Massari et al. 1986). From Late Oligocene through Early Miocene time, the kinematic front of the Dinarides shifted progressively westward into the Adriatic basin. This resulted in east-plunging downbuckling of the Venetian basin, in which less than 1 km thickness of shallow marine, Late Oligocene through

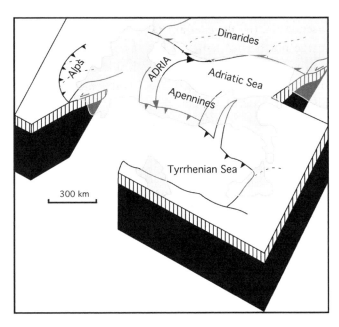

FIGURE 3.7 Block diagram perspective drawing of Adria and its relationship to adjacent plates. Adria is being subducted southwestward below the Apennines and northeastward below the Dinarides, and it has overlapped the Eurasian plate to the northwest, resulting in the southern Alps.

After Doglioni et al. 1994. Modified by permission of American Geophysical Union.

Early Miocene siliciclastic sediments accumulated as intensity of the Dinaric orogeny waned.

As Middle Miocene time began and Alpine thrusts migrated southward and were generated farther east, the Venetian basin flexed down along its northwestern margin, and a northward-thickening wedge of sediments accumulated in it during the Middle and Late Miocene (Massari et al. 1986). This wedge is over 3 km thick along its northwestern edge. Mineralogy of Late Oligocene through Early Miocene sediments suggests that they were derived from highlands around almost the entirety of the northern edge of Adria. In contrast, Middle and Late Miocene sediments were derived almost entirely from the adjacent, rising southern Alps. Downbuckling of the basin did not keep pace with sediment supply, and by Late Miocene time alluvial fans extended across the area (Massari et al. 1986). A latest Miocene (Messinian) unconformity separates the thick Miocene sequence in the basin from the later, much thinner Plio-Pleistocene sequence of intermixed marine and terrestrial deposits.

The entire northeastern side of the Adriatic Sea is bordered by the Dinarides, which formed on the overriding margin of the Eurasian plate during the Late Cretaceous to Neogene collision with Adria (e.g., Celet 1973; Miljush 1973; Aljinović et al. 1984; Picha 2002). Multiple thrust sheets, younger toward the southeast, occur along the Balkan coast of the Adriatic Sea (fig. 3.4, section C) and in general mirror the basic structure on the Italian side of the sea. In contrast with the siliciclastic-loaded sediments that constitute the Adriatic-verging Apennine thrust sheets, north of Albania the thrusts expose predominantly Upper Triassic to Lower Paleogene carbonate rocks within the internally imbricated Karst thrust sheet (Hrvatović and Pamić 2005). A narrow coastal terrane south of Dubrovnik, the Budva zone, is faulted between the stratigraphic sequences of the High Karst and Adria, and is bounded top and bottom by thrust faults (fig. 3.4, section C). A fault boundary continues northward between the High Karst and Adria north of the point at which the Budva zone pinches out. In contrast with the thick Triassic through Cretaceous carbonate platform sequences that characterize Adria and the High Karst platform, the Triassic through Cretaceous sedimentary sequence of the Budva zone is thin and consists of thin-bedded cherty limestones and radiolarian cherts, capped by graywackes and shales (Celet 1973). Sediments of the Budva zone are interpreted as remnants of the deep marine basin that originally separated Adria and the High Karst platform and was subducted as Adria moved toward Eurasia.

In the southern Adriatic basin, foredeep basins adjacent to the Dinarides-Hellenides and Apennines are separated by a relatively narrow foreland region, exposed at present as the southeasternmost geographic region of Italy known as Puglia (fig. 3.8). Adria behaves differently with respect to

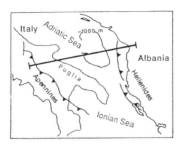

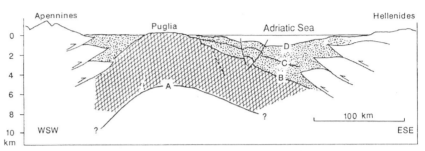

FIGURE 3.8 Structural section along a path across the southern part of Adria, from compressed southwest to compressed northeast margins, showing the double downward flexure of the near-surface parts of the plate and foredeeps between the two margins and the relatively nondeformed median foreland (Puglia).

Reprinted from *Tectonophysics*, vol. 252, de Alteriis, "Different foreland basins in Italy: Examples from the central and southern Adriatic Sea," pp. 349–373. Copyright © 1995, with permission from Elsevier.

Europe in its southern and northern portions, which are divided by the southwest–northeast line of earthquakes, the Tremiti shear zone, that crosses the plate along the southern-middle Adriatic boundary (Oldow et al. 2002). The Tremiti shear zone separates a region of thin (about 70 km–thick) lithosphere to the north and thicker (about 110 km–thick) lithosphere to the south (Doglioni et al. 1994). Doglioni et al. (1994) inferred that the thicker lithosphere of Puglia has slowed the northeastward migration of the subduction hinge in the south, while the hinge has migrated northeastward at a faster rate, but with a gentler dip, on the descending plate in the middle Adriatic region. Consequently, the Dinaride foredeep is clearly developed in the southern Adriatic basin, east of Puglia, and the Apennine foredeep occurs in the middle and northern Adriatic basin, north of the Gargano Peninsula (fig. 3.9).

In the southern Adriatic region, the transition between underlying basement rocks and the sediments that accumulated in a downbuckling foredeep is taken as the erosional unconformity that separates Cretaceous and Eocene pelagic limestones from overlying Upper Oligocene siliciclastic turbidites (de Alteriis 1995). The foredeep has continued to deepen to

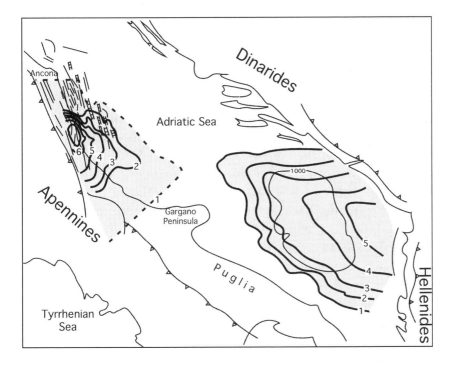

FIGURE 3.9 Depth in kilometers to the pre-Pliocene basement of the central Adriatic basin and to the pre-Oligocene basement of the southern Adriatic basin. The depth contours for the basement are drawn as heavy lines, and depths greater than 1 km are filled with gray shading. The 1,000 m isobath of the southern Adriatic basin is drawn as a thin line; note that it does not correspond to the greatest depth of the pre-Oligocene basement. The short lines within and immediately north of the central Adriatic basin fill represent north- to northwest-trending normal faults through the pre-Messinian basement and Messinian erosion surface due to downflexing of the western margin of Adria.

Reprinted from *Tectonophysics*, vol. 252, de Alteriis, "Different foreland basins in Italy: Examples from the central and southern Adriatic Sea," pp. 349–373. Copyright © 1995, with permission from Elsevier.

the present time—especially over the eastern descending edge of the plate—and to accumulate deepwater siliciclastic turbidites. The southern Adriatic basin was deep during the Late Miocene Messinian event and even then was apparently separated from the Mediterranean Sea by a relatively shallow sill across the Otranto Strait. It was one of the areas in which Messinian salts were deposited, although here they are interbedded with siliciclastic turbidites (Celet 1973; Hsü et al. 1973). Sediments were supplied to the southern Adriatic basin from westward-proliferating thrust sheets of the Dinarides and Hellenides and, eventually, also from the Apennines on the opposite (southwestern) side of the basin. Sedimentation rates reached 60–70 m Myr^{-1} during the Oligocene and Early Miocene (de Alteriis and Aiello 1993) then increased to six times the Miocene rate by Late Pliocene and remained almost that high during the Pleistocene (Bertotti et al. 2001). Nonetheless, the basin remains sediment-starved, as indicated by its 1,200 m depth and the geographic concordance of depth-to-basement and seafloor depth (fig. 3.9).

Mesozoic carbonate platform rocks that formed while Adria was located on the south coast of the Tethys Sea are overlain in the middle and northern Adriatic region by Oligocene and Early Miocene deepwater carbonate rock, on which an erosion surface developed during the Late Miocene Messinian drop in level of the Mediterranean Sea (de Alteriis 1995). The Late Miocene mid- and northern Adriatic Apennine foredeeps, however, which are now partially buried below the eastern Apennines, also contain Messinian evaporites (Rabbi and Ricci Lucchi 1966; Moruzzi and Follador 1973; Ricci Lucchi 1975; Ori et al. 1986). The evaporites thin and occur locally in discontinuous patches eastward, under the western edge of the middle Adriatic (Ori et al. 1986). Presence of Messinian evaporites in the middle Adriatic region, below the final-stage Messinian unconformity, indicates that the mid-Adriatic foredeep was deeper than the sill at the Otranto Strait, as was the southern Adriatic basin, at least early during the Messinian drop in Mediterranean sea level. In a number of localities along the Apennine foredeep, from midway along the Po Plain to the middle Adriatic, there was either continuous deposition or only a brief hiatus at the Messinian–Lower Pliocene boundary. Oligohaline bivalves and rapidly diversifying foraminifera just above the upper Messinian unconformity indicate a rapid decrease in salt content—toward normal marine salinity—as the Lower Pliocene rise in sea level began (Crescenti 1971; Esu and Taviani 1989).

The Messinian erosion surface is taken as the base of the foredeep sedimentary section in the middle and northern Adriatic region because, while there is no substantial faulting or tectonic deformation of the sediments that predates the erosion surface, the underlying sediments and the Messinian erosion surface itself were subsequently broken by normal faults as the basement was subsiding and collecting sediments on the west side of the shallow foreland (fig. 3.9). North of the Gargano Peninsula, the Apenninic foredeep is filled with a westward-thickening wedge of Plio-Pleistocene siliciclastic sediments that reaches 7 km thickness along the western shore of the middle Adriatic Sea (fig. 3.9), over 5 km thickness along the western border of the northern Adriatic Sea, and over 7 km thickness along the south edge of the Po Plain (Patacca et al. 1990). The Pliocene deposits consist of basal clay-rich turbidites that grade up-section into silty/sandy turbidites and record progressive shallowing of the basin. Pleistocene sediments are shallow-water muds to sands deposited in inclined sigmoidal beds that record delta progradation (de Alteriis 1995).

Discordance between the geographic location of the maximum sediment thickness (fig. 3.9) and the maximum depth of the middle Adriatic Sea (fig. 2.3), along with the progressively shallower environments recorded by the sediments, indicates that the mid-Adriatic basin is sediment-filled. With relatively little additional sediment accumulation, the middle Adriatic would be transformed into a low-lying sedimentary plain.

Summary

The Adriatic Sea occupies a basinal region entirely underlain by a continental microplate that originated from the southern margin of the Tethys Sea, moved north, and became attached to Europe along its northern edge. The microplate is now being subducted along at least part of its southwestern and eastern margins. Compression and overthrust faulting along the southwestern edge of the Adriatic basin has generated the Apennines, which are building progressively eastward. To the east, westward thrusting within the overriding Dinarides on the northeastern margin yields an opposing thrust-fault margin. The Apennine foredeep basins associated with the Alpine-Apennine convergence, and those associated with the more northerly approach of the Apennines and Dinarides, are represented by the terrestrial Po Plain and the very shallow northern and middle Adriatic regions. These foredeeps have post-Messinian, shallowing-upward, basin fill of Pliocene sedimentary sequences that are overlain by Pleistocene deltaic and fluvial deposits. These Pliocene and Pleistocene sediments were derived largely from the Apennines and secondarily from the southern Alps, and they prograde collectively from the western end of the Po Plain to the middle Adriatic basin.

Present-day Adriatic benthic communities live on the top surface of this sedimentary sequence, the local characteristics of which are critically important to benthic organisms that live on or in it. Types and distribution of post-Messinian through Holocene sediments are the topic of chapter 6.

The southern Adriatic basin differs in that it is a foredeep of the Dinarides, associated with the eastward subduction of Adria. Upper Oligocene to Pleistocene sediments deposited within it were derived largely from the Dinarides. The basin presently is sediment-starved and has been throughout its existence. It was apparently deeper than the sill at the Otranto Strait during the Messinian event, because marine sediment accumulation within the basin continued through the Late Miocene, and it may be the only part of the Adriatic basin that is older than Early Pliocene in age.

If the Late Oligocene through Pleistocene subduction of Adria continues, and the Apennines and Dinarides continue to converge, the northern and middle Adriatic will be transformed within a geologically short time, perhaps no more than a million years hence, into a completely terrestrial low-lying plain continuous with the Po Plain. The southern Adriatic basin will then be the recipient of progressively more Apennine-derived sediment, will become sediment-filled, and will be transformed into an extension of the Po-Adriatic plain. The Adriatic Sea will then cease to exist.

CHAPTER FOUR

PHYSICAL OCEANOGRAPHY

Ὑποπνεύσαντος δὲ νότου δόξαντες τῆς προθέσεως κεκρατηκέναι ἄραντες ἆσσον παρελέγοντο τὴν Κρήτην. μετ᾽ οὐ πολὺ δὲ ἔβαλεν κατ᾽ αὐτῆς ἄνεμος τυφωνικὸς ὁ καλούμενος Εὐρακύλων· συναρπασθέντος δὲ τοῦ πλοίου καὶ μὴ δυναμένου ἀντοφθαλμεῖν τῷ ἀνέμῳ ἐπιδόντες ἐφερόμεθα.

A south wind having sprung up, supposing that they had attained their purpose, they then weighed anchor and sailed close by Crete. But not long after there arose against it a tempestuous wind, called Euroclydon [the northeast wind]. And when the ship was caught and could not bear into the wind, we let her drive.

—*Acts of the Apostles* 27:13–15

The Adriatic Sea is, roughly, a northwest–southeast elongated, trapezoidal, semilandlocked basin, with communication with the wider ocean possible only through the small Otranto Strait at its southern end. Water flows both northward and southward through the strait, but given that the Adriatic is a net exporter of water, more water flows south into the Mediterranean than enters through the strait. Surface circulation patterns are of more immediate importance to the northern Adriatic benthic ecosystem than is the deepwater circulation. However, surface circulation and deepwater circulation are interlinked in that the nutrient balance—and the functioning of the ecosystem— for the sea as a whole depends ultimately on the way the integrated system operates. ^{90}Sr is an isotope of strontium that was abundant in radioactive fallout up to the 1960s, when a moratorium on testing came into effect, and it is a good radioactive tracer of mass transport of marine water. Based on flow rates at the Otranto Strait and residence time of ^{90}Sr within the Adriatic, the entire system is tightly integrated with turnover of water between the Adriatic and the Mediterranean of between one and four years (Franić 2005).

SURFACE CIRCULATION

Circulation of Adriatic surface water is most strongly influenced by the interplay of three component forces (Artegiani et al. 1997b). First is the inflow of fresh water from point sources, especially the Po River, which individually accounts for over a quarter of the freshwater runoff into the Adriatic. Second is inflow of Mediterranean water through the Otranto Strait. And the third is wind shear, which is temporally variable and also differs in overall pattern between the northern and southern portions of the Adriatic (Cavaleri et al. 1997).

BAROCLINIC-DRIVEN CYCLONIC FLOW

The Po River flows into the Adriatic at an average rate of 1,600 m^3 sec^{-1} (Raicich 1994), entering roughly mid-way along the irregular northwest edge of the trapezoidal basin. In the absence of any other forces, this input of low-density fresh water would generate a dynamically unstable, tographically high wedge subject to density-driven, down-slope flow (baroclinic flow). Gravitational flow down the surface of a freshwater wedge is deflected to the right in the northern hemisphere.

Such a flow pattern should generate a current that arcs southward from the Po Delta, meets the Italian coast, and, because of continued deflection to the right, continues as a southward-flowing current along the Italian coast to and through the western side of the Otranto Strait. Modeling of the baroclinic flow generated by discharge from the Po River develops just such a flow pattern, although the local bulge in the coastal profile caused by the Po Delta generates northward leakage and ponding of fresh water along the coast as well (fig. 4.1A). Similarly, water from all the other, lesser Adriatic rivers (fig. 2.2) should be deflected to the right (fig. 4.1B). Even without other influences, freshwater flow into the Adriatic should produce an overall cyclonic (counterclockwise) flow. Consistent with this expectation, salinity maps averaging a forty-year composite data set show lowest surface salinity in the Adriatic in the vicinity of the Po Delta, with seasonably variable southward extension of a low-salinity wedge along the Italian coast (fig. 4.2).

Net gain of fresh water over evaporation of the Adriatic Sea (see chapter 2) causes water to be exported into the Mediterranean through the Otranto Strait. There is also a long-known counterflow of Mediterranean water from the Ionian Sea into the Adriatic (fig. 4.3; Zorè 1956; Buljan and Zore-Armada 1976). This Mediterranean water flowing north through the Otranto Strait is affected by the same rightward deflection as is the water entering the Adriatic from rivers. In enclosed estuaries and seas of the northern hemisphere, this rightward deflection ("Coriolis effect") interacts with the margins of such basins to generate cyclonic flow, so the incoming Aegean water tends to flow northward along the eastern Adriatic coast. This flow is particularly clear in the autumn, when the inflowing warm

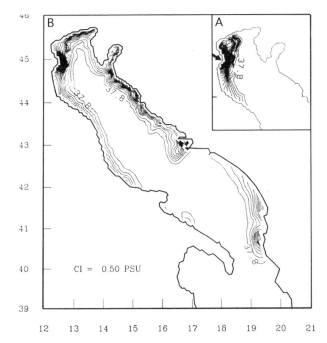

FIGURE 4.1 Modeled surface-sa-
linity distribution within the Adriatic
thirty days after initially "turning
on" freshwater flow down rivers entering the sea, based on (A) buoyancy
forcing by the Po River only and (B)
buoyancy forcing by the Po and the
other rivers shown in fig. 2.2.

Fig. 4.1A reprinted from *Journal of Marine Systems*, vol. 30, Kourafalou, "River
plume development in semi-enclosed
Mediterranean regions: North Adriatic
Sea and Northwestern Aegean Sea," pp.
181–205. Copyright © 2001, with permission from Elsevier. Fig. 4.1B after
Kourafalou 1999; reproduced by permission of American Geophysical Union.

(>17°C) Mediterranean water is largely confined to the eastern coast (Artegiani et al. 1997b). The general pattern of cyclonic flow within the Adriatic is virtually ubiquitous, with the northward flow off the Istrian coast
(fig. 2.3) continuing into the Gulf of Trieste, where water below about 10 m
depth maintains cyclonic flow of 2–3 cm sec^{-1} (Stravisi 1983).

Cyclonic surface circulation of Adriatic water has been known since the
earliest comprehensive studies of water movement in the sea (e.g., Zorè
1956; reviewed in Orlić et al. 1992). Analyses of decades-long data (Artegiani et al. 1997b) and shorter-term satellite tracking of transmission from
drifters (Poulain 1999, 2001) confirm the pattern (fig. 4.4). The surface circulation is best interpreted as baroclinic flow driven primarily by two
forces, inflow of fresh water from the Po River and of Mediterranean water
through the Otranto Strait, supplemented by freshwater input from the
secondary Adriatic rivers.

Coastal configurations and, especially, the deep southern Adriatic basin,
the basin-transverse Pelagosa Sill, the middle Adriatic depression, and the
transition from the middle Adriatic depression to the northern Adriatic
shelf break up the overall basinwide flow into three cyclonic gyres (fig. 4.4).
The south Adriatic gyre is permanently centered over the southern basin.
The middle Adriatic gyre is centered over the middle Adriatic depression
during three seasons but is somewhat variable in position during the summer (Zorè 1956; Mosetti and Lavenia 1969; Malanotte-Rizzoli and Bergamasco 1983). The northern Adriatic gyre, located over the deeper part of

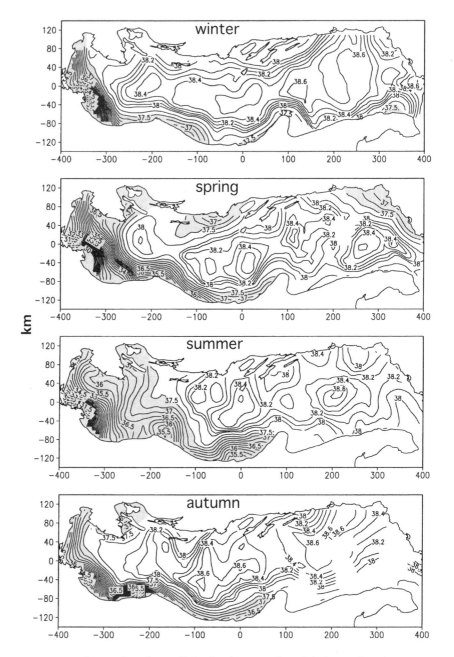

FIGURE 4.2 Seasonal surface salinity (psu) across the Adriatic Sea, based on composite data from 5,543 bottle-cast stations during 1911–1914 and 1947–1983 (from Artegiani et al. 1997b). "Spring" includes May–June data, "autumn" includes November–December data, and the other two seasons are each defined as four months; these uneven divisions were used because of the relatively rapid transitions between winter and summer states. Regions of the sea where the surface is diluted more than 1 psu by freshwater runoff are approximated by the 37 psu contours, and the areas with 37.5 psu surface water are shaded. Note the concentration of decreasing-salinity contours toward the Po Delta in all seasons, and the extent of low-salinity water (shaded) southward along the Italian coast.

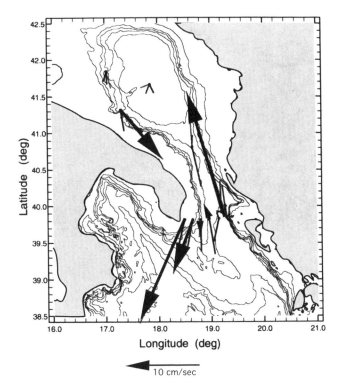

FIGURE 4.3 Nineteen-month (May 1994–November 1995) average vectors for surface currents from six bottom-tethered current meters arrayed across the Otranto Strait. Northward flow of Mediterranean water into the Adriatic is concentrated along the eastern coast, and southward flow of Adriatic surface water into the Mediterranean is concentrated along the western coast.

Reprinted from *Journal of Marine Systems*, vol. 20, Kovačević et al., "Eulerian current measurements in the Strait of Otranto and in the southern Adriatic," pp. 255–278. Copyright © 1999, with permission from Elsevier.

the northern Adriatic shelf, is seasonal, complex, and highly variable (Artegiani et al. 1997b), and there is a seasonally developed fourth gyre in the shallow, northern end of the northern Adriatic (fig. 4.4B, C). In fact, there are seasonal variations in intensity of surface currents throughout the Adriatic Sea, although the largest variation in both intensity and direction of currents is found in the shallow northern Adriatic.

WIND-DRIVEN FLOW

Much of the variation in surface circulation is due to wind direction, speed, and duration. Wind speed is highest throughout the entire basin from late November through January, during this time temporarily reaching over 5 m sec^{-1} as a daily average in each of the three regions with instantaneous speeds over 30 m sec^{-1}; it is least during May and June (Cavaleri et al. 1997). Prevailing winds within the Adriatic as a whole are the "bora" (Croatian: "bura"), which is the name given to wind from the northeast, and the "sirocco" (Croatian: "jugo"), which is the name given to wind from the southeast. The bora blows across the Adriatic and is therefore an offshore wind for the eastern coast and onshore for the western coast, and modeling results in low sea-surface elevation along the Balkan coast and high surface

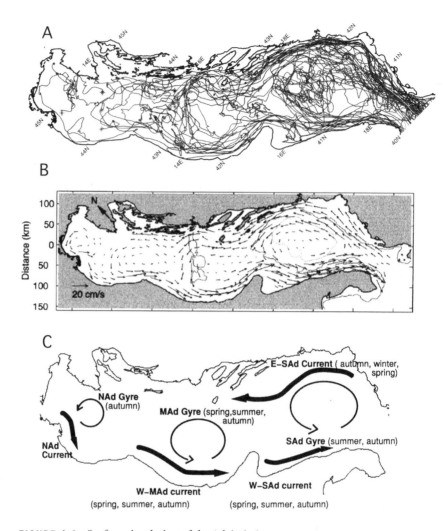

FIGURE 4.4 Surface circulation of the Adriatic Sea.

A) Trajectories of drifters released in six batches between December 1994 and October 1995 and observed at six-hour intervals December 1994 through March 1996; release points marked by stars and final good transmission sites marked by solid circles.

Reprinted from *Journal of Marine Systems*, vol. 20, Poulain, "Drifter observations of surface circulation in the Adriatic Sea between December 1994 and March 1996," pp. 231–253. Copyright © 1999, with permission from Elsevier.

B) Average direction and velocity of surface flow throughout the Adriatic as determined from drifters tracked from August 1990–July 1999.

Reprinted from *Journal of Marine Systems*, vol. 29, Poulain, "Adriatic Sea surface circulation as derived from drifter data between 1990 and 1999," pp. 3–32. Copyright © 2001, with permission from Elsevier.

C) Annual circulation determined from forty years of data.

From Artegiani et al. 1997b.

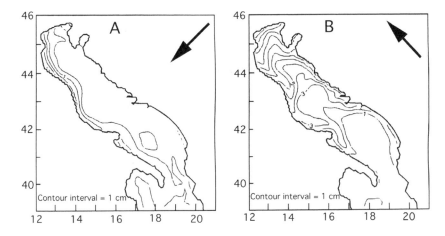

FIGURE 4.5 Modeled surface elevation of the Adriatic Sea resulting solely from stress-forcing by 10 m sec^{-1} winds at the end of five days for the A) bora and B) sirocco. Five days is a typical duration for windstorms in the Adriatic region.

Reprinted from *Journal of Marine Systems*, vol. 30, Kourafalou, "River plume development in semi-enclosed Mediterranean regions: North Adriatic Sea and Northwestern Aegean Sea," pp. 181–205. Copyright © 2001, with permission from Elsevier.

elevation along the Italian coast (fig. 4.5A), especially in the northwest in the vicinity of Venice (Bergamasco and Gačić 1996). The sirocco blows along the length of the Adriatic, and modeling results in high surface elevation in the northern regions, with maximum elevations against the northwest coast (fig. 4.5B). Note that either wind tends to push water toward Venice (fig. 2.7).

In the northern Adriatic the bora is the prevailing wind, whereas the bora and the sirocco are nearly co-dominant in the middle and southern Adriatic (Cavaleri et al. 1997). The pattern of storms with winds in excess of 10 m sec^{-1} is somewhat different, however. Well over half the windstorms in the northern Adriatic consist of the bora; the remaining storms in the region are more-or-less random with respect to compass direction, except that few come from the northwest because of the presence of the Alps (Cavaleri et al. 1997). While the most common type of windstorm in the middle Adriatic consists of the sirocco, there is overall greater heterogeneity in storm wind direction; well over half the storms in the southern Adriatic consist of the sirocco and slightly more easterly winds (Cavaleri et al. 1997). Overall there is a continuous increase in the proportion of mild and storm sirocco from north to south, at least along the Croatian coast (Lisac et al. 1998), and a concomitant rapid decline in frequency of the bora.

High wind events can disrupt even the essential surface-flow patterns of the Adriatic Sea. The outflow of excess fresh water through the Otranto

Strait averages approximately 3,700 m³ sec⁻¹, based on Raicich's (1996) esti-mate of the annual freshwater balance, and this excess fresh water is mixed with and slightly reduces the overall salinity of an even larger volume of sea water that flows out of the Otranto Strait. The total exchange through the strait varies from year to year (Manca et al. 2002) but is estimated to aver-age about 0.4 Sverdrup (1 Sv = 10⁶ m³ sec⁻¹), with lower summer and higher winter surface flow (Michelato and Kovačević 1991). Despite the continu-ous force generated by the outflow, exceptionally strong sirocco can cause surface flow to reverse so that it flows northward down to at least 50 m deep along the Italian coast of the strait (e.g., Michelato and Kovačević 1991; Kovacevic et al. 1999).

Similarly, wind- and barometric pressure–induced reversals can result in southward flow along the Albanian coast of the strait (Bergamasco and Gačić 1996). Even the usually persistent western Adriatic current can be reversed in the middle and northern Adriatic by a combination of unusually strong si-rocco and substantially reduced freshwater input from the Po River (Pou-lain et al. 2004). Such reversals are most common in the summer, the season of weakest surface-water inflow from the Mediterranean (Michelato and Kovačević 1991; Kovačević et al. 1999; Poulain 1999). Reverse-flow events are ephemeral, lasting essentially only as long as the causative wind exists.

Each of the major elements of the surface circulation (fig. 4.4C) of the Adriatic is seasonally variable (table 4.1). The seasonal variations are most pronounced in the northern Adriatic because of its shallow depth and con-sequent greater perturbation by density changes associated with seasonal variation in fresh water, intensity of evaporation and temperature changes, and wind-generated currents that can quickly reach to the seafloor.

CIRCULATION IN THE NORTHERN ADRIATIC

General circulation in the northern Adriatic region is complex, varying profoundly between seasons and often behaving quite differently from year to year. The northern Adriatic behaves largely as an open sea during fall and winter. In these seasons, vertical instability throughout the water col-umn, along with strong and frequent bora storms, promotes vigorous cir-culation that receives water from the mid-Adriatic along the east coast and sends water back into it along the west coast (e.g., Mosetti and Lavenia 1969; Zore-Armanda and Gačić 1987). During the spring and summer, rel-atively low-density water of the northern Adriatic interchanges less readily with waters to the south, and the area behaves oceanographically as a semi-enclosed sea (Degobbis 1989; Krajcar 2003). Movement of water within the northern Adriatic therefore is both density-driven and wind-driven, and sorting out the relative influence of the various forcing mechanisms

TABLE 4.1 Seasonal Strength and Seasonal Attributes of the Major Elements of the Surface Circulation of the Adriatic Sea

	Spring	Summer	Autumn	Winter
South Adriatic				
Eastern current	+	+++	+++	+−
Wind	v	v	NE, SE, V	NE, V
Runoff	+	+−	++	++
Western current	+	++	++	+−
Wind	v	v	NE, SE, V	NE, V
Runoff	+−	+−	+−	+−
Gyre	+	++	+++	+−
Wind	v	v	NE, SE, V	NE, V
Middle Adriatic				
Western current	+++	+−	+++	+
Wind	se	v	NE, SE, V	SE, V
Runoff	+−	+−	+−	+−
Gyre	+	++	+++	+−
Wind	se	v	NE, SE, V	SE, V
North Adriatic				
Current (west coast)	+++	+−	+++	+
Wind	v	ne	NE	NE
Runoff	++	+−	+++	+
Gyre	+−	+−	++	+−
Wind	v	ne	NE	NE

Key for flow strength: +++ = strong, ++ = moderate; + = weak; +− = weak or absent.

Key for winds: NE = bora; SE = sirocco; V = variable directions

Capital letters indicate two or more storms per month, lowercase letters indicate fewer than one storm per month over a twenty-nine-year period of observation.

These generalized summaries of relative intensity are based on data from Artegiani et al. 1997b for circulation; from Raicich 1994 for local surface runoff; and from Cavaleri et al. 1997 for wind strength.

requires an understanding of the effects of salinity, temperature, and density stratification.

WINTER CIRCULATION

Distribution of mountain masses and lowlands in the northern Balkan Peninsula locally concentrates the bora so that it varies in speed from place to place where it blows off land and onto the Adriatic. Higher wind speeds are achieved immediately northwest and southeast of the Istrian Peninsula than occur off the peninsula itself (fig. 4.6), and that difference is most pronounced during autumn and winter (fig. 4.7).

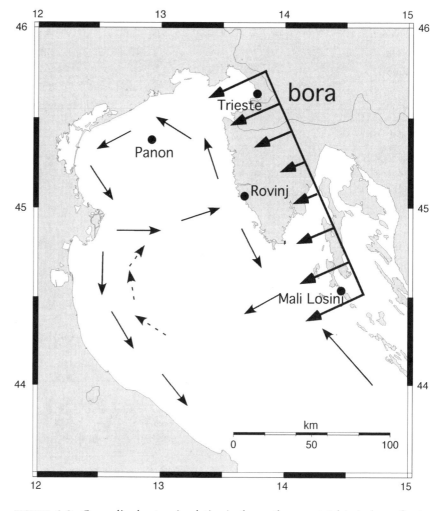

FIGURE 4.6   Generalized water circulation in the northernmost Adriatic due to forcing by differential velocity of the bora (after Zore-Armanda and Gačić 1987).

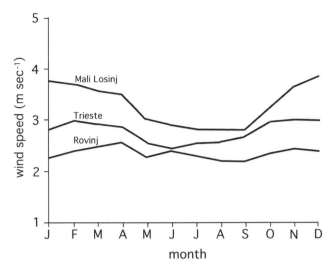

FIGURE 4.7 Average monthly wind speed in m sec⁻¹ for 1966–1992 at Trieste, Rovinj, and Mali Losinj (after Supić and Orlić 1999); see locations in fig. 4.6. Note that there is a greater difference in wind speed between Rovinj and the other two localities from October through April than during spring and summer.

The northern band of strong bora is a primary driving force for the cyclonic gyre centered north of the Po Delta and Rovinj, Croatia, and the southern band of strong bora helps to drive the general Adriatic cyclonic circulation while at times setting up a weak anticyclonic gyre (fig. 4.6) between the cyclonic cells (e.g., Zore-Armanda and Gačić 1987; Orlić et al. 1994; Mauri and Poulain 2001; Paklar et al. 2001; Krajcar 2003). The anticyclonic gyre does not extend to the Italian coast because, in addition to setting up the two northern Adriatic gyres, the bora also moves water onto the Italian coast, producing a topographic high along shore (fig. 4.5A), which enhances the baroclinic coastal flow generated by the Po and other northern Italian rivers (fig. 4.1B).

Consequently, the winter northern Adriatic current south of the Po Delta is closely confined to the Italian coast and has a sharply defined, steep boundary with the more saline open-basin waters (fig. 4.8A). This pattern of open exchange with the more southerly parts of the Adriatic is reflected in winter contrasts in salinity on opposite shores of the northern Adriatic (fig. 4.9). Northward flow of high-salinity water along the Balkan coast results in high salinity of water offshore of Rovinj, on the Istrian coast, corresponding with the lowest annual salinity of water at Fano, Italy, due to vigorous winter export of Po water southward along the Italian coast (Orlić 1989).

However, pronounced currents in the northern Adriatic require continual forcing because of frictional retardation by the shallow seafloor. The bora does not blow continuously, and the other forces that generate the overall cyclonic circulation farther south in the Adriatic are not very effective in the northern Adriatic because it is so shallow. Response to strong winds in the northern Adriatic is immediate, generating strong downwind flow north

A. Early December 1978

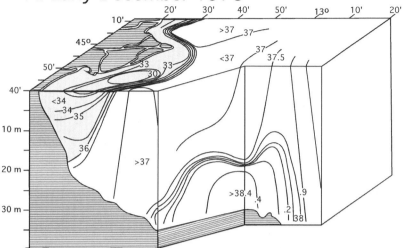

B. Early July 1978

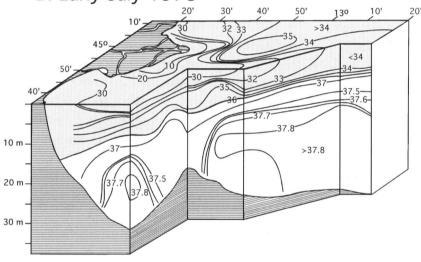

FIGURE 4.8 Distribution of salinity in the northern Adriatic offshore of the Po Delta. Shaded, <36 psu areas approximate the river-diluted coastal waters, and the open-basin water has higher salinity.

A) Early winter coastal confinement of the northern Adriatic current.

B) Midsummer basinward spread of river-diluted water over denser, higher-salinity open-basin water.

Reprinted from *Science of the Total Environment*, supplement 1992, Franco and Michelato, "Northern Adriatic Sea: Oceanography of the basin proper and of the western coastal zone," pp. 35–62. Copyright © 1992, with permission from Elsevier.

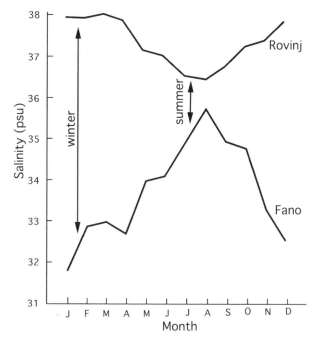

FIGURE 4.9 Comparison of annual patterns of surface salinity at Fano, on the Italian coast south of the Po Delta, and Rovinj, on the Istrian coast opposite the Po River.

From Orlić 1989.

and south of Istria and upwind flow off the peninsula where the bora is relatively weaker (Orlić et al. 1994; Lee et al. 2005); the storm-induced current ceases within a few hours after the end of the storm (fig. 4.10).

This generalized understanding has been well documented by satellite radiometry, by which a newly established cold, low-salinity filament extending into the open sea from the Italian coast can be detected following strong bora events (fig. 4.11). Combined satellite images of surface chlorophyll *a* and continuous tracking of surface drifters has demonstrated intermittent existence of a two-gyre system correlative with bora storms even as early as September and October (Mauri and Poulain 2001). In summary, although even some eastern coastal areas are characterized by a persistent flow direction (Brana and Krajcar 1995), the northern winter circulation gyres are ephemeral, forced by an interaction of density of the water, which is determined by both salinity and temperature, and wind shear (Franco and Michelato 1992; Paklar et al. 2001).

From September through May, the sirocco constitutes approximately 10 percent of the northern Adriatic windstorms (Cavaleri et al. 1997). Sirocco storms result in high sea level, with a pronounced slope up to the northwestern coast (fig. 4.5B), and acqua alta in Venice associated with strong sirocco is well known back to the nineteenth century (Orlić et al. 1994). Flow across the northern–middle Adriatic boundary area, modeled on an actual wind field during a strong sirocco, yielded upwind (northward) flow across almost the entire area. Even the north Adriatic current

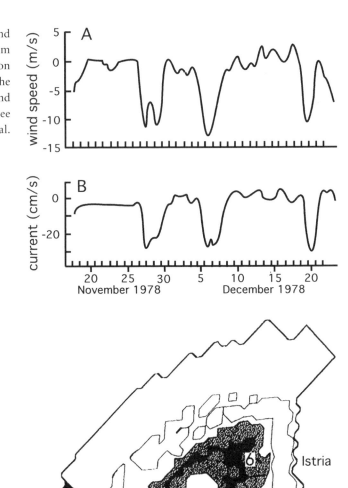

FIGURE 4.10 East-northeast wind speed (A) and current speed at 25 m depth (B) recorded at the Panon offshore station (fig. 4.6) in the northern Adriatic in November and December 1978, spanning three bursts of the bora (from Orlić et al. 1986).

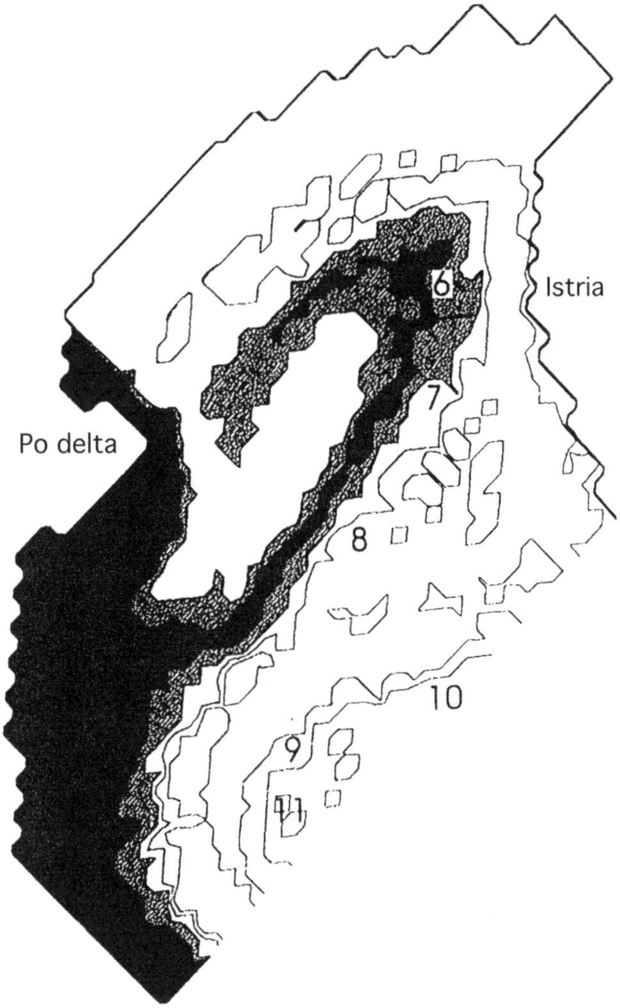

FIGURE 4.11 Satellite radiometer–determined temperatures (°C) following a bora windstorm in 1987, showing an eastward filament of cold water from the Po Delta toward Rovinj, then curving north and west in a short-lived cyclonic current.

Reprinted from *Continental Shelf Research*, vol. 21, Paklar et al., "A case study of bora-driven flow and density changes on the Adriatic Shelf (January 1987)," pp. 1751–1783. Copyright © 2001, with permission from Elsevier.

(NAC) can be reduced and in some instances reversed by a strong sirocco (Orlić et al. 1994). Such severe reduction in strength of the NAC results in ponding of the freshwater runoff from the Po and other northern Italian rivers, with a wedge of freshening waters extending progressively farther across the northern Adriatic (fig. 4.12C).

SUMMER CIRCULATION

All the circulation-promoting forces that cause the shallow northern Adriatic to function as an open body, exchanging water with the other parts of the sea during the winter, are weakened or are absent during the summer. There are no summertime compensating forces.

Rate of river input into the Adriatic influences circulation rate, both overall and locally, with higher input driving circulation more vigorously (Kourafalou 1999, 2001). Summer is the period of lowest discharge for the Po River (under 75 percent of the annual mean flow rate) and of the total discharge into the Adriatic (under 60 percent of the annual mean) (Raicich 1994).

Aside from gentle, onshore breezes, winds are usually quiescent over the northern Adriatic during the summer. Only about 8 percent of the total windstorms with wind speeds over 10 m sec^{-1} occurred during June–August in contrast with about 40 percent during December–February for a twenty-nine-year data set (Cavaleri et al. 1997).

During the winter the middle of the northern region is the coldest part of the Adriatic, which results in southward export of dense, cold water. (See below for the generation and fate of cold, dense water from the northern part of the sea.) During the summer and early autumn, however, the warm, low-density water in the region (Artegiani et al. 1997b) does not sink into the deeper, more southerly regions.

With the decline of circulation-generating forces as summer approaches, the northern Adriatic becomes almost isolated from interaction with the other regions of the sea. Lack of vigorous circulation is evident in the paths of drifters released in May 1995 (fig. 4.13), although collectively they do indicate a general cyclonic circulation and the existence of a weak gyre confined to the northeastern end.

In the absence of vigorous southward flow of the NAC, fresh water from the Po and other northern Italian rivers mixes with the northern Adriatic water to form one or more surface layers of progressively more diluted water that extend farther and farther toward the eastern coast (fig. 4.8B). The summer nearconfinement of the Po water within the northern Adriatic is clearly indicated by the summer increase in salinity off Fano, Italy, and corresponding decrease in salinity off Rovinj, Croatia. Salinity at the two

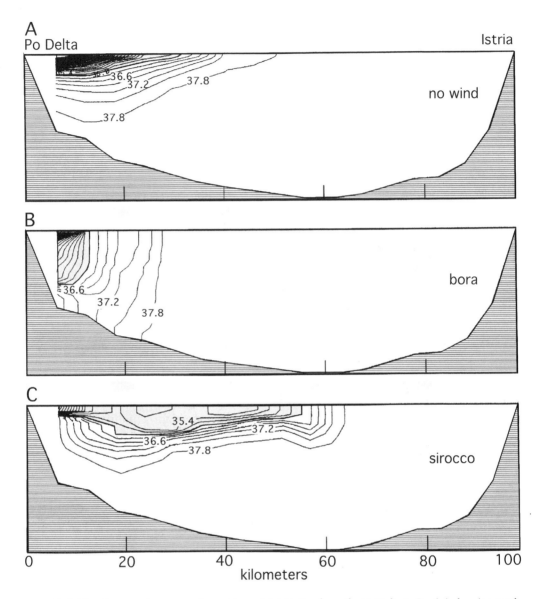

FIGURE 4.12   Cross section across the northern Adriatic Sea from the Po Delta to Rovinj, showing modeled salinity distributions 30–35 days after "turning on" the Adriatic rivers with (A) no wind, (B) uniform bora wind from the northeast at 10 m sec^{-1} and (C) sirocco wind from the southeast at 10 m sec^{-1}.

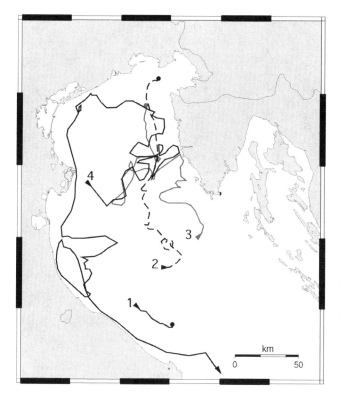

FIGURE 4.13 Trajectories of satellite-tracked drifters released in the northern Adriatic in early May 1995. Arrows mark the point of release, and solid circles mark the point of last transmission. Lifetime of the drifters: 5 days (#1), 31 days (#2), 37 days (#3), and 79 days (#4). After leaving the northern Adriatic, drifter #4 continued traveling south along the Italian coast until transmission ceased midway between Ancona and the Gargano Peninsula.

Reprinted from *Journal of Marine Systems*, vol. 20, Poulain, "Drifter observations of surface circulation in the Adriatic Sea between December 1994 and March 1996," pp. 231–253. Copyright © 1999, with permission from Elsevier.

coastal localities can become almost identical in August, due to the surficial pool of Po-diluted water that spreads across the entire northern Adriatic (fig. 4.9), developing a weak cyclonic pattern that can be due exclusively to thermal properties (Bergamasco et al. 2003). The tongue of low-salinity water spreading from the Po Delta also contributes to the weak, ephemeral late-summer anticyclonic gyre south of the Po–Rovinj line (Brana and Krajcar 1995; Supić et al. 2000; Krajcar 2003).

Even with the development of stratification, substantial warming occurs all the way to the shallow seafloor throughout the northern Adriatic, yielding an average temperature that is 4°C higher even at 75 m—a depth reached only just north of the northern–middle Adriatic boundary—than at the same depth in the middle and southern parts of the sea (Artegiani et al. 1997b). Both low salinity and high temperature reduce the density of northern Adriatic water during summer and autumn, further distinguishing the region and reflecting its relative isolation from interaction with the rest of the sea.

Variation in water temperature can interact with the other variables to cause substantial year-to-year fluctuations in summer circulation (Orlić 1989). The profound effect of summer temperature can be seen in plots of Po discharge and salinity off Rovinj for a two-year period (fig. 4.14). The

FIGURE 4.14 (A) Po River outflow and (B) surface salinity at Fano, on the Italian coast south of the Po Delta, and Rovinj, on the Istrian coast (from Orlić 1989). The annual summer decrease in salinity at Rovinj and increase at Fano (fig. 4.9) were so great in 1958 that salinity at Rovinj crossed below salinity at Fano. This was due apparently to reduced density of northern Adriatic water because of anomalously high summer temperatures such that the northern Adriatic current stopped and low-salinity water ponded across the entire width of the northern Adriatic.

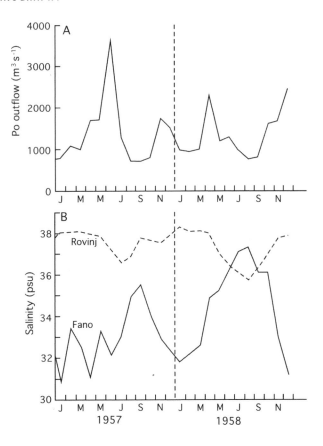

very high spring maximum in Po discharge for 1957 corresponds with a clear decrease in salinity offshore of Rovinj, but in 1958 the salinity at Rovinj was deeply depressed even though there was a smaller spring outflow from the Po than in the preceding year. The northern Adriatic sea surface temperatures in spring 1958 were about 2°C higher than average. Apparently, the low-density water caused by the high temperature inhibited late spring and summer circulation even more than usual, resulting in more intense pooling of the Po water within the surface waters of the northern Adriatic and thereby depressing salinity even along the east coast.

Interaction of wind and flow rate of the Po River can cause excursions from the normal seasonal patterns. During the summer, protracted high discharge from the Po River, in the absence of winds exceeding 5 m sec⁻¹, results in the spread of Po waters across to the Istrian Peninsula as a low-density surface layer. However, summer flow less than 1,000 m³ sec⁻¹ for as little as two weeks, still in the absence of vigorous winds, reduces the Po plume to a narrow strip confined to the Italian coast (Sturm et al. 1992). Winter conditions apparently are more variable: of ten satellite images during January through April 1982, five spectral analyses indicate water from the Po extending well across the basin, four indicate confinement of the Po plume alone the Italian coast, and one was intermediate (Sturm et al. 1992). Three of the analyses followed periods of intermediate (1,000–1,500 m sec^{-1}) discharge of the Po, which resulted in coastal confinement of its plume except following a strong bora event. During times that the discharge was low, weak winds resulted in coastal confinement of the plume, and strong winds (both bora and sirocco) spread the Po waters eastward (e.g., fig. 4.11).

In summary, summer and winter circulation patterns in the northern Adriatic are on average quite different from one another. A single pattern of eastward-spreading low-density water from the Po River and the absence of well-developed cyclonic circulation predominates during the summer. During the winter cyclonic circulation is much more common, regulated by strong bora storms. Several studies independently have noted an apparently well-established cyclonic gyre north of the line from the Po Delta to Rovinj, and in fact an extension of fine sediments as a narrow band curving toward the northeast (fig. 6.7) from the Po Delta requires the existence of that confined gyre.

SUBSURFACE CIRCULATION

High surface-water temperatures during summer in the northern Adriatic contribute importantly to its summer isolation from the rest of the Adriatic. During summer a thermocline exists at 10–30 m depth in the northeastern Adriatic, with surface water temperatures rising to 22–25°C (Buljan and Zore-Armanda 1976). During winter there is no thermocline, temperatures fall to 11–12°C, and the entire water column becomes unstable across most of the northern Adriatic, that is, it can freely overturn from top to bottom. This water does not remain confined to the northern Adriatic. Instead, it flows slowly down-slope as dense, cold, winter water, gradually mixing with other water further south to produce an important influence on Adriatic and eastern Mediterranean deepwater circulation.

SUBSURFACE FLOW THROUGH OTRANTO STRAIT

The 800 m sill depth of the Otranto Strait is well below the depth at which low-density surface waters flow from the Adriatic into the Mediterranean along the Italian coast, with contrary flow into the Adriatic along the Albanian coast. Four water masses pass through the Otranto Strait (fig. 4.15). The two surficial streams are referred to as the Adriatic surface water (ASW) and Ionian surface water (ISW), respectively. (The Ionian Sea is that part of the Mediterranean adjacent to the strait, the origin of the surficial

FIGURE 4.15 Salinity profile across the Otranto Strait, in which can be seen the four water masses that constitute the exchange between the Adriatic and Mediterranean seas (after Artegiani et al. 1997a).

ADW = Adriatic deep water (<38.5 psu); ASW = Adriatic surface water (<38.3 psu); ISW = Ionian surface water (<38.3 psu); LIW = Levantine intermediate water (>38.6 psu).

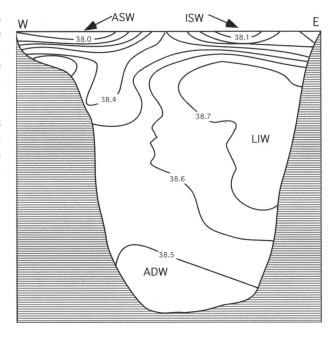

northerly flow into the Adriatic.) Water from the eastern Mediterranean (Levantine intermediate water, or LIW) moves into the Adriatic below the two surficial flows, and Adriatic deep water (ADW) flows over the sill and into the Mediterranean. Very few actual data on currents through the strait were available until near the end of the twentieth century, but the water masses involved were known much earlier (e.g., Pollak 1951).

The LIW is focused on the eastern slope of the Otranto Strait, centered at about 400 m depth. Although concentrated at the bottom of the strait, the ADW also extends somewhat up along the western slope so that there is an oblique boundary between the two subsurface masses. Some mixing occurs because of shear that generates slow-moving eddies between the two. However, the pronounced, shear-generated eddying—which characterizes the two surface streams—is not characteristic of the deeper, more stable subsurface currents as they pass through the strait (Michelato and Kovačević 1991). The subsurface currents do vary in lateral extent and velocity from season to season and even over periods of a few days. These deepwater current variations depend on discharge of the Po and other rivers and on atmospheric forcing (Poulain et al. 1996; Manca et al. 2002), and there can be huge variations in the mass of water involved from year to year (Vilibić and Orlić 2002). Even short-term events such as reversal of high and low atmospheric pressure between the northern Adriatic and the Ionian Sea affect the flow rate of the LIW, and flow rate of the ADW increases to compensate for sirocco reversal of the ASW flow (Michelato and Kovačević 1991).

LEVANTINE INTERMEDIATE WATER

LIW enters the Adriatic as a warm, highly saline (13.5–14.5°C, >38.8 psu) stream that extends from approximately 250 to 500 m depth (Artegiani et al. 1997a; Wu and Haines 1998). From that point the LIW flows in a cyclonic path within the southern Adriatic basin. Once within the Adriatic, the LIW becomes modified by advection and mixing but is recognizable within the southern Adriatic as a high-salinity mass (>38.6 psu) between approximately 100 m and 400 m depth, except in the winter (fig. 4.16F). LIW can also be recognized as an even further diluted mass (>38.5) in the middle Adriatic at somewhat shallower depths, approximately 50 to 150 m, except in the winter (fig. 4.16E). There is no trace of it in the northern Adriatic salinity profile (fig. 4.16D).

The maximum exchange rate through the Otranto Strait likely occurs in the late winter or spring, immediately following a December–January minimum (Wu and Haines 1998). Based on an array of Eulerian and Lagangrian current measurements across the Otranto Strait, Poulain et al. (1996) estimated total inflow into the Adriatic through the strait to be 1.67±0.26 Sv (1 Sv = 10^6 m^3 sec^{-1}) during February 1995, 0.45±0.16 Sv of which consisted of LIW, dropping to 0.65±0.27 Sv in May 1995, 0.14±0.27 Sv of which consisted of LIW. This variation in the intra-annual flow is the probable reason for the winter absence of a distinct high-salinity mass at the appropriate depths for the LIW in the averaged salinity profiles of the southern and middle Adriatic (fig. 4.16E, F).

ORIGIN OF ADRIATIC DEEP WATER

ADW is an essential component generating the deepwater mass across much of the eastern Mediterranean (Pollak 1951; Roether and Schlizer 1991), where its high density is the result of low temperature despite its relatively low salinity as it leaves the Otranto Strait (Artegiani et al. 1997a). Relative to deep water of the middle and northern portions, deep water in the southern Adriatic is warmer and more saline (table 4.2), which suggests that ADW is the result of mixing of LIW and Adriatic-derived surface water (Artegiani et al. 1997a).

The progressive increase in both temperature and salinity of deep water from the northern to the southern Adriatic, and the northward shallowing and decreased salinity of the LIW (fig. 4.16), indicates that advection and mixing of Adriatic and Levantine water masses do indeed occur as they move in opposite directions along the length of the Adriatic. Water temperatures below 12°C occur at the surface only during winter, and the northern Adriatic is the only region where such low temperatures are the

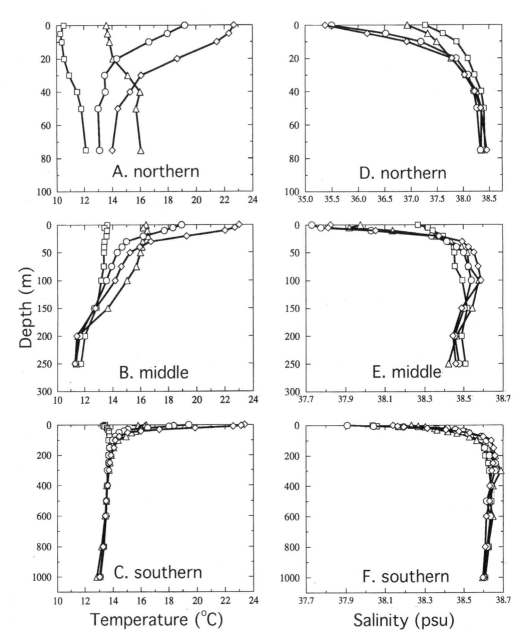

FIGURE 4.16 Seasonal temperature (A–C) and salinity (D–F) profiles averaged from several decades of data for the northern, middle, and southern Adriatic (after Artegiani et al. 1997a). Symbols for seasons are identical for each graph: winter (square), spring (circle), summer (diamond), fall (triangle). LIW is indicated spring through autumn by the highest-salinity water in the southern Adriatic between 100 and 400 m depth and in the middle Adriatic between 50 and 150 m depth; it is absent in the northern Adriatic.

TABLE 4.2   Regionally Averaged Temperature, Salinity, and Density of Adriatic Deep Water

Region	Temperature (°C)	Salinity (psu)	Density (sigma–t)
Northern Adriatic	11.35 ± 1.40	38.30 ± 0.28	>29.2
Middle Adriatic	11.62 ± 0.75	38.47 ± 0.15	>29.2
Southern Adriatic	13.16 ± 0.30	38.61 ± 0.09	>29.1

From data compiled by Artegiani et al. (1997a).

average (fig. 4.16A–C). Temperature profiles, along with the gradual warming and increased salt content of the ADW to the south, point to the northern Adriatic as an important influence on the ADW, even though the northern Adriatic deep water (NAdDW) constitutes a very small proportion of ADW in the southern Adriatic basin (Gačic et al. 1999; Vilibić and Orlić 2002).

High-density water is formed in the northern Adriatic, especially in shallow coastal areas from the Gulf of Trieste to the vicinity of Venice, in association with several months of reduced flow from the Po River and the cold, dry, winter bora storm events (Mosetti and Lavenia 1969; Supić and Orlić 1999; Vilibić and Supić 2003, 2005). Strong bora winds cause both a pronounced decrease in temperature of surface waters and increased salinity because of high evaporation rates. Very high density water (sigma-t >29.5) flows southward along the western seafloor at the boundary between the northern and middle Adriatic. Some is then lost down slope and into the middle Adriatic depression, where it mixes with local water to become warmer and more saline (Artegiani et al. 1997a; Vilibić et al. 2004).

Continued movement of recognizable NAdDW southward along the Italian coast suggests cyclonic flow for bottom waters as well as for surface waters in the northern Adriatic, but with a strong density boundary between the relatively fresh NAC and the saltier NAdDW (Franco and Michelato 1992; Artegiani et al. 1997a). In spring middle Adriatic deep water flows over the Pelagruza Sill and into the southern Adriatic, and a recognizable strand of NAdDW is commonly recognizable along the Italian coast as far south as Bari, south of the Gargano Peninsula, where it cascades down and mixes into the ADW (Manca and Giorgetti 1999; Vilibić and Orlić 2002).

Strong bora storms of several days' duration in January and February cause sufficient turbulent mixing and rapid cooling to result in vertical instability of the water column from surface to seafloor through most of the northern Adriatic, which feeds development and southward flow of the NAdDW. There is a compensating generalized northward flow into the northern Adriatic of warmer, more saline water extending from the surface to the seafloor along the Croatian coast (Brana and Krajcar 1995; Artegiani et al. 1997a). Dense-water formation is actually highly variable spatially and temporally throughout the winter, and under conditions of low runoff from the Po River and extreme bora storms, pools of dense water can even form along the Croatian coast of the northern Adriatic (Supić and Orlić 1999).

SUMMARY AND IMPLICATIONS

The broad-scale circulation within the Adriatic Sea results in very different qualities of water along the east and west coasts. Warm, moderately high-salinity Ionian surface water enters the Otranto Strait along the Albanian coast and flows northward along the Balkan coast. Very few rivers of consequence flow from the Balkans into the Adriatic, so the north-flowing water offshore has been only slightly reduced in salinity at the point where it flows into the northern Adriatic. Deeper in the sea, the high-salinity Levantine intermediate water enters via the Otranto Strait and flows northward only as far as the middle Adriatic.

The Po River and the majority of other important rivers are located along the north and western coast of the northern Adriatic. Their waters mix with open Adriatic water to produce a low-salinity stream that follows the Italian coast southward to the Otranto Strait. The northern Adriatic is so shallow that, during the late summer or early autumn, its waters become so warm or relatively fresh that interchange with the rest of the Adriatic becomes very slow and can essentially stop altogether. When this happens, the relatively fresh water that characterizes the Italian coast can spread briefly as far as Istria, at the northern end of the Balkan coast. During winter, in addition to the surficial north Adriatic current, water also exits the northern Adriatic as a very cold bottom flow that becomes focused as a slow-moving western subsurface stream. This subsurface stream ultimately descends into the southern basin and mixes with Levantine intermediate water to drive the Adriatic deepwater current that flows south over the sill of the Otranto Strait and spreads as a deep, cold layer through the western Mediterranean.

Along the east coast of the Adriatic, the chemistry of shallow water is therefore largely determined by the chemistry of the ISW. In contrast, composition of shallow water along the west coast is determined by the proportions at which the ISW intermixes with fresh water introduced via the rivers entering along that coast. In addition to causing a reduction in salinity, fresh water has a very different concentration and ratio of nutrient ions than does marine water. This has fundamentally important ecological effects. Therefore the circulation system determines that there are important differences in the pelagic and ultimately the benthic marine ecosystems off the Balkan and the Italian coasts. These differences extend into the northern Adriatic, as described in the following chapters.

CHAPTER FIVE

NUTRIENTS AND PELAGIC BIOLOGY

The [Dalmatian] water was hardly water, being fused with sunshine.
—REBECCA WEST, *Black Lamb and Grey Falcon*

Nutrient availability, the rate at which nutrients are taken into primary producers, and the paths and rates at which nutrients are transferred from one organism to another determine the structure of an ecosystem. In shallow, sunlit waters, pelagic and benthic photosynthetic organisms are in competition with one another and also with nonphotosynthetic bacteria for the available nutrients. Both the overall concentration and the relative availability of each of the basic nutrients—inorganic nitrogen, phosphorus, carbon—along with micronutrient and oxygen availability influence which of the competing organisms are most successful. The differential success of the various primary producers and their total rate of production have profound consequences that determine the structure and stability of the benthic ecosystem. It has been suggested that an increase in nutrient availability in the world's oceans through geological time has been important in restructuring benthic ecology, at least at shelf depths (Bambach 1993, 1999; Martin 1996; Allmon and Ross 2001). Relationship of the Adriatic ecosystem to this hypothesis will be taken up in chapter 9.

In this chapter, the sources and seasonal fluctuations in concentration of the primary nutrients in the Adriatic are discussed, followed by an assessment of the fertility of the northern Adriatic relative to the more southerly parts of the Adriatic and the Mediterranean in general, and then an overview of the relationship between nutrients and the pelagic ecosystem of the northern Adriatic. First, however, a few definitions are in order.

DETERMINATION OF TROPHIC LEVELS

Quantification is important in any consideration of nutrient availability and its flow through an ecosystem. But quantification of what?

- Which nutrients? The three basic nutrients are inorganic compounds of carbon (C), nitrogen (N), and phosphorus (P), plus compounds of silicon (Si) for diatoms or others with silicate skeletons.
- Instantaneous concentration of individual nutrients within the water? This is generally given as $\mu mol\ l^{-1}$ or $\mu mol\ dm^{-3}$, or less frequently $mmol\ m^{-3}$.
- Rate of introduction of nutrients, via rivers, upwelling, and/or recycling? A large variety of metrics, usually per day or per year, is used for this.
- Rate at which primary producers incorporate nutrients? This is generally given as $\mu mol\ l^{-1}\ t^{-1}$ or $mmol\ m^{-3}\ t^{-1}$, or—if the entire water column or the sediment surface is taken as a unit—$mmol\ m^{-2}\ t^{-1}$ (t^{-1} is some unit of time, generally day or year).
- The amount of biomass in primary producers and/or consumers at any given time? This is known as *standing crop* and is given as $g\ C\ m^{-3}$ or, if the entire water column is totaled, $g\ C\ m^{-2}$.
- The rate at which biomass increases? This is known as *productivity* and is generally given as $g\ C\ m^{-3}\ t^{-1}$ or $g\ C\ m^{-2}\ t^{-1}$ (t^{-1} being per hour, day, or year).

All these various measures give useful information, and although measures of rate are on average more informative than are measures of mass at a given instant, measures of rate are harder to obtain. All the measures commonly vary from season to season and on shorter predictable (day versus night) and unpredictable scales. The scale and timing of the variation are themselves important in understanding ecosystem dynamics, but complete time series usually are lacking. With the advent of remote sensing from satellites, the entire *surface* of the sea can be monitored for instantaneous reflective color, which can be adjusted for regional conditions and used to estimate the mass of chlorophyll *a* per liter or per m^3 in surface water. This estimate can then be used to approximate mass of chlorophyll *a* per m^2 based on assumptions about the rate at which the concentration changes with depth.

Regardless of what measure is used, comparison of results from one point to another indicate that, in the sea as on land and in fresh water, there are areas that are rich—maybe even too rich—in nutrients and organisms, in contrast with areas that are biological deserts. Marine biologists have adopted from limnologists a set of terms that describe this variation: *oligo-trophic* water is relatively a biological desert, *eutrophic* water is rich, and *mesotrophic* water is intermediate between the two end members. There is an unfortunate tendency, however, to use these as undefined terms, without specifying even approximate boundaries, and to base them on different attributes as listed in the previous paragraph.

DEFINITIONS

Nixon (1995) was apparently the first to suggest a comprehensive trophic classification scheme for coastal marine ecosystems, taking his cue from Rodhe's 1969 proposed trophic classification scheme for fresh waters that was based on organic carbon input measured in g C m^{-2}. His scheme is based on organic carbon entering the biological system:

oligotrophic <100 g C m^{-2} yr^{-1}
mesotrophic $100-300$ g C m^{-2} yr^{-1}
eutrophic $301-500$ g C m^{-2} yr^{-1}
hypertrophic >500 g C m^{-2} yr^{-1}

Note that these terms are adjectives that modify the often unspecified noun "system," and they refer to total carbon supplied to a system, whether it be "fixation by primary producers within the system of concern (autochthonous carbon) or an input of organic matter from outside the system (allochthonous carbon)" (Nixon 1995:201). The widely used noun "eutrophication," related to this scheme, was defined by Nixon (1995:199) as "an increase in the rate of supply of organic matter to an ecosystem."

The appeal of Nixon's (1995) scheme for trophic classification of coastal marine ecosystems is its simplicity, even though it doesn't take into account shorter-term variations in productivity such as seasonal algal blooms alternating with periods of extremely low growth of algae within the water column. More complex trophic indices have been proposed that are based on several aspects of water quality. For example, Vollenweider et al. (1998) proposed an index derived from their study of the northwestern Adriatic coastal waters, based on standing crop (chlorophyll *a*)—a proxy for rate of proliferation of algae—and availability of the primary nutrients nitrogen and phosphorus. Using their trophic index, Vollenweider et al. (1998) demonstrated that water offshore of the Emilia-Romagna coast was less productive than in an onshore zone, that productivity of the onshore zone decreases southward away from the Po River's nutrient influx, that the trophic state fluctuates in a patterned way during the year, and that it is closely tied to rate of immediate discharge rate of the Po. However, such detailed analysis is not easily scaled up to large areas at present without extensive analysis of data from numerous sources as well as new, systematic acquisition of data. In this book, carbon fixation and its surrogate chlorophyll *a* concentration (see below) are used as indicators of trophic state instead of more complex formulations.

APPLICATIONS

At present annual productivity in the open Atlantic Ocean averages from 120 g C m^{-2} in the Trade Wind belts, through 142 g C m^{-2} in the Westerlies domain, to 373 g C m^{-2} in open polar waters (Sathyendranath et al. 1995). Using Nixon's definitions for trophic conditions, two of the open-ocean domains (Westerlies and Trades) are near the low range of mesotrophic, whereas the polar domain is eutrophic. However, within the Westerlies and Trades domains, some regions (south Atlantic subtropical gyre, western portion of the north Atlantic subtropical gyre) are characterized by <100 g C m^{-2} yr^{-1} (Sathyendranath et al. 1995) and thus are oligotrophic according to Nixon's scheme. Inasmuch as the north and south Atlantic subtropical gyres have long been characterized as oligotrophic, it seems reasonable to apply Nixon's classification scheme to the open ocean basins as well as to coastal regions. Sathyendranath et al. (1995) found annual coastal productivity to be well above 300 g C m^{-2} in each broadly defined coastal region. They also reported biomass in 0–50 m deep coastal waters, based on chlorophyll *a* determination, to range from about 1 mg C m^{-3} in fall and winter to over 3 mg C m^{-3} in spring. In contrast, the Trade Wind domains have mean values increasing from 0.3 to 0.6 mg C m^{-3} from spring through winter (Sathyendranath et al. 1995).

Many studies report data on carbon in the northern Adriatic biomass as mg chlorophyll *a* m^{-3}. Based on a rough approximation of the correspondence between g C m^{-2} yr^{-1} and mg chlorophyll *a* m^{-3} from Sathyendranath et al. (1995), a second operational set of definitions of trophic terms is used in this book:

oligotrophic < ~0.5 mg chlorophyll *a* m^{-3}
mesotrophic ~0.5 – ~1.5 mg chlorophyll *a* m^{-3}
eutrophic   ~1.51 – ~2.5 mg chlorophyll *a* m^{-3}
hypertrophic > ~2.5 mg chlorophyll *a* m^{-3}

In almost all instances use of the terms "oligotrophic," etc., in reference to the Adriatic will be based on mg chlorophyll *a* m^{-3}, but whether Nixon's scheme or the operational definitions are used in any given instance can be determined from the units of the data used.

There is a more local study that justifies use of the chlorophyll *a* standing crop as a proxy for productivity in g C m^{-2} yr^{-1}. In a repeatedly occupied transect across the coastal Po plume and into open Adriatic water, Zoppini et al. (1995) found a high correlation (r = 0.91) between chlorophyll *a* and primary production, with the relationship

Primary Productivity = 0.16 + 2.84 chlorophyll *a*

where rate of primary production is expressed in mg C m^{-3} hour^{-1} and chlorophyll *a* concentration is expressed in mg m^{-3}. In the absence of more extensive comparisons within the northern Adriatic, this relationship is used as the basis for estimating primary production from chlorophyll *a* data in other studies.

Several methods have been used to determine productivity and standing crop in the sea, and they vary in precision and completeness. For example, picoplankton (plankton <2 μm diameter) generally were not included in determination of chlorophyll *a* in the northern Adriatic prior to 1988 (Zoppini et al. 1995), because of the pore size in the filters used. However, as noted by Zavatarelli et al. (1998:233) in an analysis of data collected by various measures across forty-seven years, "Apparently, the inhomogeneities due to the different methods of analysis are not significant, perhaps because they are masked by the high variability in the properties concentration." In other words, given the rather large range in concentrations encountered, systematic differences in data from before 1988 and after 1988 are swamped, and the earlier and later data can be considered of equal value for characterizing the nutrient distributional pattern of the northern Adriatic.

NUTRIENTS REQUIRED BY PHYTOPLANKTON

Major nutrients required for the growth of phytoplankton are inorganic compounds of carbon (C), nitrogen (N), phosphorus (P), and—at least for phytoplankton such as diatoms that build silicate skeletons—inorganic compounds of silicon (Si). In the open ocean, the ratios required for C:N:P are 106:16:1 (Redfield et al. 1963), the "Redfield ratio." Where Si is required, the N:Si ratio is essentially 1:1 (Richards 1958). In the absence of other influences, a 106:16:15:1 ratio for dissolved inorganic C:N:Si:P within surficial marine water would be in nutritive balance for growth of diatom-rich phytoplankton.

Carbon is an order of magnitude in excess in marine water in relation to the other nutrients. The amount of carbon fixed into a marine ecosystem is controlled by the availability of the other nutrients, and its availability therefore is not discussed below. Any of the other basic nutrients that has a smaller ratio relative to the others would be used up first and therefore would be the limiting nutrient for the system. However, there may be differences among the nutrients in regeneration rate from preexisting organic matter, differential rate of input from an external source, or adaptation by primary producers to use a lower proportion of a perennially severely limited nutrient. The result might be that the ratio of nutrients in a coastal water mass may differ from the Redfield ratio yet still not give a completely accurate insight into nutrient dynamics.

DISTRIBUTION OF PRIMARY NUTRIENTS

In parallel with the compilation of a synthesis on the general circulation of the Adriatic Sea (see the beginning of chapter 4 above; Artegiani et al. 1997a, b), a multi-institutional Italian CNR team compiled biogeochemical data collected from open waters of the Adriatic through the second half of the twentieth century and generated a comprehensive analysis of the essential distribution of basic nutrients and oxygen (Zavatarelli et al. 1998). Much of the description below is based on this compilation, which divided the northern Adriatic into two subregions, the shallow northern Adriatic northwest of the 40 m isobath (essentially northwest of a line from Rimini to Rovinj) and the deep northern Adriatic southeast of the 40 m isobath (fig. 2.3).

Nutrients in the shallow northwestern Adriatic generally have high surface concentrations, decrease downward to depths of about 5–10 m, and are uniform or increase in concentration below 10 m (Zavatarelli et al. 1998). The high concentration of nutrients at the sea surface reflects their influx via low-density water from Italian rivers (see below), which do not flow into the deeper northern Adriatic (and where a similar surficial spike in nutrient concentration is not found).

NITROGEN

Total dissolved nitrogen is present in water as nitrate (NO_3), nitrite (NO_2), and ammonium (NH_3), but many studies of nitrogen concentration report only nitrate. In the shallow northern Adriatic, nitrate concentration is on average highest throughout the water column during the winter, although the concentration at the sea surface spikes (7 µmol l^{-1}) in spring, and nitrogen is almost completely depleted in the summer except just above the seafloor (Zavatarelli et al. 1998). In contrast, in the deep northern Adriatic, nitrate barely exceeds 1 µmol l^{-1} throughout the water column from autumn through spring and virtually disappears in the surface layer down to 30 m during summer.

Surface distribution maps of nitrate for each season, plus winter and summer profiles for the shallow and the more southerly, deeper northern Adriatic are given in figure 5.1. The Rimini to Pula transect represents the deep northern Adriatic as a whole, for this and the other nutrients, because the values and patterns seen along this path are essentially identical in a transect from Ancona to Zadar, near the boundary between the northern and central Adriatic (Zavatarelli et al. 1998). In general, winter nitrate values diminish eastward across the western and middle part of the shallow northern Adriatic, whereas high values are limited to the shallow western coast waters in the Rimini to Pula transect. Nitrate is only minimally present at depths less than 20 m in both transects for summer but is moderate at depth and increases toward the seafloor. Faganeli and Herndl (1991) found a significant relationship between dissolved organic nitrogen and bacterial biomass in the Gulf of Trieste, suggesting that nitrogen is an important limiting nutrient for bacterioplankton in the gulf.

As can be inferred from figure 5.1, the Po River is the major source of nitrate entering the Adriatic. The annual load of nitrate added via the Po increased from under 50 million kg yr^{-1} in the late 1960s to 100 million kg yr^{-1} in the early 1990s (de Wit and Bendoricchio 2001).

Particulate organic nitrogen (PON) also enters the Adriatic with Po River water. From early 1995 through early 1996, PON was added at a rate of 19.4×10^3 tonnes per year via the Po, fluctuating through the year in parallel with particulate organic carbon (Pettine et al. 1998). PON within the Adriatic is very high during the stratified summer season, twice as high above the pycnocline as below (68 vs. 23 μg l^{-1}), and overall as much as an order of magnitude higher than during winter (Gilmartin and Revelante 1991). Not all PON in the Adriatic is flushed down rivers; the summer high of PON is at least in part due to organic particles formed within the sea itself.

PHOSPHORUS

Total dissolved phosphorus is present in water as orthophosphate (PO_4) and dissolved organic phosphorus compounds, but most studies of phosphorus concentration report only (ortho)phosphate, because it is much more abundant than the other forms. Phosphate reaches its maximum average concentration in the shallow northern Adriatic during the autumn and is minimum during winter and spring, with maximum autumn average of 0.15 μmol l^{-1} at 20 m depth and minimum spring average of 0.04 μmol l^{-1} at 5–10 m depth. In contrast, average phosphate concentration in

Nitrate

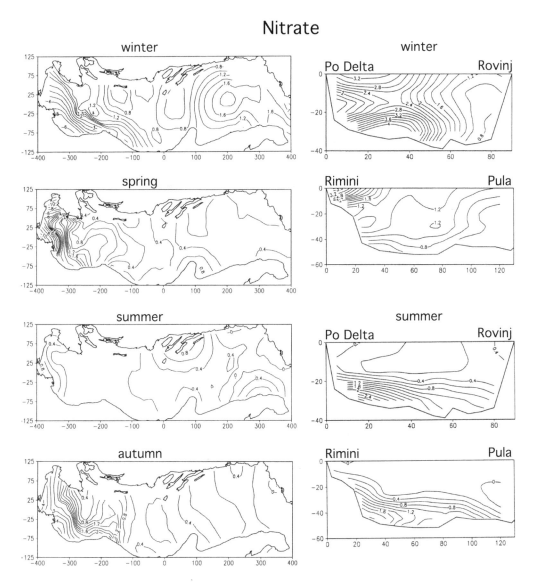

FIGURE 5.1 Surface distribution maps of nitrate in μmol l⁻¹ for each season plus winter and summer profiles for the shallow (Po Delta to Rovinj) and the deep (Rimini to Pula) northern Adriatic, based on data from 1948 to 1991.

Reprinted from *Journal of Marine Systems*, vol. 18, Zavatarelli et al., "Climatological biogeochemical characteristics of the Adriatic Sea," pp. 227–263. Copyright © 1998, with permission from Elsevier.

the deeper northern Adriatic is highest during the winter, 0.05–0.10 μmol l^{-1} between the surface and 50 m depth, and 0.27 μmol l^{-1} at 75 m depth; the other three seasons have rather uniform vertical profiles with values at or below 0.05 μmol l^{-1} (Zavatarelli et. al. 1998).

Surface distribution maps of phosphate for each season plus winter and summer profiles for the shallow and the deep northern Adriatic are given in figure 5.2. In general, winter phosphate values are highest in surface water midway across the shallow northern Adriatic, whereas the distribution is more complex in the Rimini to Pula transect, with lowest values throughout the water column in the east. In the summer there is a steep gradient in phosphate concentration along the Italian coast in both transects, with highest values at the seafloor about 20–30 m deep.

The Po River is the primary source of phosphorus in the Adriatic. The annual load of total phosphorus delivered per year by the Po peaked in 1979–1981 at just over 14 million kg yr^{-1} and was down to about 8 million kg yr^{-1} by the early 1990s (de Wit and Bendoricchio 2001). Seasonal variations in phosphate are about four times greater near the Po River than in the eastern northern Adriatic (Revelante and Gilmartin 1976).

SILICON

Silicon is present in water almost exclusively as orthosilicic acid (H_4SiO_4), although in the oceanographic literature the term "orthosilicate" and symbol SiO_4 are generally used. Orthosilicate reaches its maximum average concentration in the upper 10 m of the shallow northern Adriatic during the summer and autumn and is minimum during winter (Zavatarelli et al. 1998). In contrast, average silicate concentration down to 10 m over the deeper northern Adriatic is relatively high during the autumn and winter and lowest during the summer, although there is little difference in concentration between seasons in this region.

Surface distribution maps of orthosilicate for each season plus winter and summer profiles for the shallow and the deep northern Adriatic are given in figure 5.3. In general, as seen in the winter profiles, orthosilicate values are uniformly distributed vertically throughout the water column and are highest in the eastern part of the shallow northern Adriatic, whereas the distribution is more uniform horizontally but somewhat more stratified in the Rimini to Pula transect. During the summer throughout the northern Adriatic, silicate is depleted because of diatom proliferation down to the main pycnocline (about 10 m in the northern transect and 20 m deep in the southern transect), immediately overlying a steep gradient of increase to a deep region of high concentrations that continue to increase right down to the seafloor. A study along the 20 m depth contour south of

Phosphate

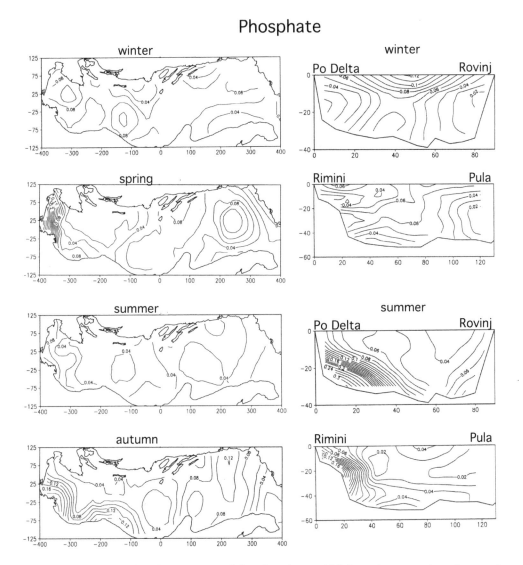

FIGURE 5.2 Surface distribution maps of phosphate in μmol l⁻¹ for each season plus winter and summer profiles for the shallow (Po Delta to Rovinj) and the deep (Rimini to Pula) northern Adriatic, based on data from 1948 to 1991.

Reprinted from *Journal of Marine Systems*, vol. 18, Zavatarelli et al., "Climatological biogeochemical characteristics of the Adriatic Sea," pp. 227–263. Copyright © 1998, with permission from Elsevier.

Silicate

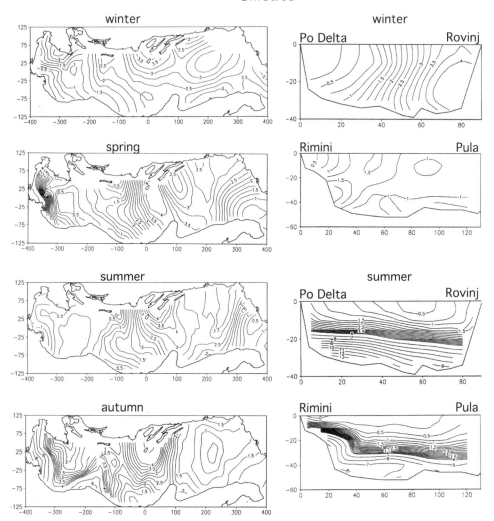

FIGURE 5.3 Surface distribution maps of silicate in μmol l⁻¹ for each season plus winter and summer profiles for the shallow (Po Delta to Rovinj) and the deep (Rimini to Pula) northern Adriatic, based on data from 1948 to 1991.

Reprinted from *Journal of Marine Systems*, vol. 18, Zavatarelli et al., "Climatological biogeochemical characteristics of the Adriatic Sea," pp. 227–263. Copyright © 1998, with permission from Elsevier.

the Po Delta found orthosilicate evenly distributed from the surface to 16 m depth, but with as much as 10^3 increase in silicate concentration at 10 cm above the seafloor (Giordani et al. 1997).

CARBON

Carbon is required in higher proportions than any of the other nutrients, but it is seldom if ever limiting. The ratio of carbon to nitrogen atoms required by marine organisms is 204:16 (~6.6:1), but the ratio of carbonate carbon to nitrate nitrogen in marine waters typically exceeds that by two times (Redfield et al. 1963: fig. 3). Most inorganic carbon assimilated by phytoplankton and benthic plants eventually is available as particulate organic carbon (POC) within the water, eventually consumed by bacteria and released as dissolved organic carbon (DOC).

Inorganic carbon is added to marine water from terrestrial flow, submarine dissolution of calcium carbonate in shells and carbonate rocks, and by addition of carbon dioxide from the atmosphere. Aside from drainage into the Po basin from its southern flank of the Apennines, the Po and all other rivers entering the northern Adriatic almost exclusively drain carbonate terrains consisting of limestones and dolostones. These rivers are saturated or nearly saturated with CO_2, largely as bicarbonate (HCO_3^-).

Organic carbon is added from river water and by decomposition of marine organisms. Particulate organic carbon is very high in the northern Adriatic during the stratified summer season, when it is twice as high above the pycnocline as below (446 vs. 210 µg l^{-1}), and overall it is as much as an order of magnitude higher than during winter (Gilmartin and Revelante 1991). Much of this comes down the Po River, which in the mid-1990s contributed 14.6×10^4 tonnes POC plus 13.4×10^4 tonnes DOC per year (Pettine et al. 1998). The DOC has much the same annual pattern in the northern Adriatic as does POC, although with much lower variation in concentration. It is about twice as high in June (~150 µmol l^{-1}) as in February (~80 µmol l^{-1}) and is concentrated in low-density surface waters in summer but is uniformly distributed through the water column in winter (Pettine et al. 1999). The lower winter concentration apparently is due to flushing by the winter circulation pattern.

ANNUAL CYCLICITY

As in most bodies of water, there is an annual cycle of peaks in dissolved organic nutrients in the northern Adriatic. Within the Gulf of Trieste, DOC peaks during the summer, dissolved organic nitrogen peaks in summer and in winter, and dissolved organic phosphorous peaks in the spring. "The observed variations . . . are controlled principally biologically through (1) release of DOM [dissolved organic matter] by phytoplankton and benthic macrophytes during the various stages of their growth and decay, and release from detrital material, and (2) by DOM utilization by heterotrophic organisms" (Faganeli and Herndl 1991:60).

GEOCHEMICAL BUDGET OF NUTRIENTS

Geochemical overviews are difficult to do for complex regions such as the northern Adriatic, even though in comparison with the major ocean basins it is rather small. Consequently all geochemical statements or conclusions are imprecise to some extent. Geochemists often are happy with order-of-magnitude approximations, especially when inferences are based on the interaction of several variables. Generally, the larger the data set on which geochemical overviews are based, the greater the accuracy of the inferences. However, for many questions valuable insights can be gained even from data that yield order-of-magnitude approximations. This gave Degobbis and Gilmartin (1990) courage to examine the biogeochemical flux of nitrogen, phosphorus, and silicon in the northern Adriatic Sea—based on about 20,000 marine analyses plus published information on chemistry of coastal discharges.

The primary nutrient inputs into near-shore marine areas are river flow and other types of discharge from the land, such as underwater springs and cumulative human waste discharge; introduction via inflowing marine water; and introduction from the atmosphere through the sea surface. The primary avenues of loss are by water flowing out of the area, sedimentation, alteration to an unusable (refractory) state, and, in some regions, extraction by land-based predators, notably human fisheries. Figure 5.4 summarizes the annual flux of the primary nutrients through the northern Adriatic as approximated by Degobbis and Gilmartin (1990). Values for gains and losses differ at most by 35 percent (silicon), which suggests that estimates are within an order of magnitude of accuracy and therefore can provide some insight.

The nitrogen budget (fig. 5.4A) demonstrates that Italian rivers are by far the largest source for inorganic nitrogen, and that rainfall and cyclonic flow of water from farther south in the Adriatic contribute an approximately equal amount of inorganic nitrogen. However, input of total nitrogen—inorganic plus organic, given in brackets in figure 5.4—from the Italian coast and from the cyclonic flow along the east coast are approximately equal. The Italian coastal sources are predominantly anthropogenic wastes, but the organic matter introduced in the eastern sector is

A. Nitrogen

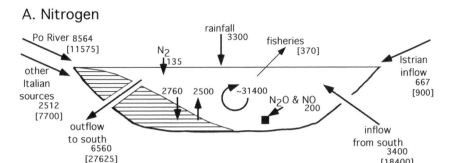

Po River 8564 [11575]

rainfall 3300

fisheries [370]

N₂ 135

Istrian inflow 667 [900]

other Italian sources 2512 [7700]

2760 2500 ~31400

N₂O & NO 200

outflow to south 6560 [27625]

inflow from south 3400 [18400]

B. Phosphorus

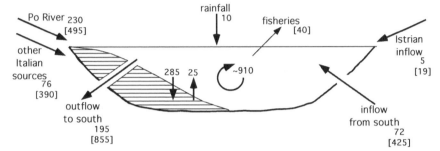

Po River 230 [495]

rainfall 10

fisheries [40]

Istrian inflow 5 [19]

other Italian sources 76 [390]

285 25 ~910

outflow to south 195 [855]

inflow from south 72 [425]

C. Silicon

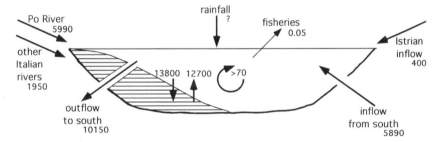

Po River 5990

rainfall ?

fisheries 0.05

Istrian inflow 400

other Italian rivers 1950

13800 12700 >70

outflow to south 10150

inflow from south 5890

FIGURE 5.4 Overall biogeochemical budgets for nutrients in the northern Adriatic Sea. Values are in 10^6 mol y^{-1}.

A) Total inorganic nitrogen (NO_3, NO_2, NH_4) and for total nitrogen including inorganic nitrogen, dissolved organic nitrogen, and particulate organic nitrogen. Total inorganic nitrogen values are not enclosed, but total nitrogen values are given in parentheses.

B) Phosphate and total phosphorus [bracket numbers], including both inorganic and organic forms.

C) Dissolved silicon [as $Si(OH)_4$].

Based on data in Degobbis and Gilmartin 1990.

largely from normal operation of the pelagic ecosystem in more southerly waters.

The water that flows southward out of the northern Adriatic, along the Italian coast, carries almost twice as much inorganic nitrogen and 50 percent more total nitrogen than does the water that flows into the northern Adriatic along its eastern coast. A relatively trivial amount of nitrogen (1 percent of the total) is permanently lost into the sediments. About 30 percent more nitrogen is regenerated within the water column each year than is added to the northern Adriatic from external sources. This within-water-column regeneration, and also that of phosphorus, is the result of grazing and production of waste products by zooplankton and of decomposition by bacteria. The bottom layer of sea water and the sediment–water interface appear to be the most vigorous sites of microbial transformation of organic particles to dissolved nutrients (Faganeli and Herndl 1991).

Over 25 percent of the total phosphate is lost into the sediment (fig. 5.4B), which contrasts with the small loss of nitrogen into the same sink. Otherwise, the pattern for phosphorus is essentially the same as for nitrogen, though involving a much smaller total.

A high proportion (~90 percent) of the silicon (fig. 5.4C) in the system is deposited on the seafloor but is almost matched by the amount that returns to the water column (i.e., is recycled, mainly from diatom frustules) at or near the sediment surface. Only about 8 percent of the silicon is actually lost into the sediment each year. Therefore, the sediment–water interface is the most important boundary for both loss and gain of silicon for the northern Adriatic. This is reflected in the silicate concentration fields seen in the profiles in figure 5.3. Setting aside the vigorous interaction at the sediment–water interface, the silicon flux through the northern Adriatic generally resembles that of the other two nutrients.

The approximately equal contribution of new nutrients and recycling by regeneration in the northern Adriatic is unusual, even for estuaries and other coastal areas, which typically have proportionally much smaller external sources (Degobbis and Gilmartin 1990). The high external contribution appears to be the sole reason that the northern Adriatic has higher productivity overall than do the middle and southern parts of the sea.

NUTRIENT LIMITATION

Phosphorus is inferred to be the principal limiting nutrient in the northern Adriatic during late winter–spring blooms, particularly in areas where sea and fresh waters mix (e.g., Chiaudani et al. 1980; Zoppini et al. 1995; Zavatarelli et al. 2000). Overall, however, the basinwide N:P ratio is rather variable, with averages in the northern Adriatic of 17:1 in shallow water and 20:1 in deeper water, based on a 1948–1991 data set (Zavatarelli et al. 2000). These averages are fairly close to the Redfield ratio. With the intermixture—and fluctuating distribution—of various water masses in the northern Adriatic, whether P or N is the limiting nutrient is spatially and temporally variable, with N often being the limiting nutrient in the oligotrophic waters flowing into the northern Adriatic from farther south (e.g., Smodlaka 1986; Degobbis et al. 2000). Severe depletion of orthosilicate in surface water during the summer (fig. 5.3, Rimini–Pula transect) suggests that despite Si having almost twice the concentration of N in the northern Adriatic as a whole (Zavatarelli et al. 2000), under certain conditions it too may be a limiting nutrient for diatom growth.

Phytoplankton in the northern Adriatic may have been ecologically selected to be efficient primary producers in the presence of N:P ratios higher than the 16:1 that generally is optimum. Various lines of evidence are consistent with this hypothesis: diatoms are capable of reproduction in phosphorus-deficient media, utilizing all available nutrients even up to ratios of 30:1, and N:P ratios within northern Adriatic zooplankton tissues (27:1) and their excretions (18:1) suggest that they feed on phytoplankton with N:P ratios higher than the Redfield ratio (Degobbis 1990).

TROPHIC STATE

Overall the northern Adriatic Sea is oligotrophic in the eastern areas and mesotrophic in the western part, with eutrophic zones off the Po Delta and in the near-coastal belt of the Emilia-Romagna region, south of the Po Delta. Chlorophyll a concentration is about 1 mg m^{-3} spring through autumn to a little over 2 mg m^{-3} in winter for the shallow northern part, and close to 0.5 mg m^{-3} in spring and summer and slightly higher during autumn and winter in the deeper northern Adriatic (Zavatarelli et al. 1998).

As seen in figure 5.4, inorganic nutrient introduction in the northern Adriatic is overwhelmingly via the Po River (Degobbis and Gilmartin 1990; Gilmartin and Revelante 1991). The Adriatic-wide circulation system sweeps the higher-nutrient waters southward along the Italian coast during the winter (Artegiani et al. 1997b; Degobbis et al. 2000). This southerly transport of nutrients out of the northern Adriatic is less effective during spring and summer, allowing a considerable part of the low-salinity, nutrient-enriched water to spread to the east coast by late summer or autumn (Brana and Krajcar 1995; Krajkar 2003), which may result in benthic anoxia (e.g., Stachowitsch 1984, 1991; Justić 1991; Hrs-Brenko et al. 1994). Overall, however, the cyclonic flow is effective at removing the riverine nutrients so that the northern Adriatic, while on average nutrient-rich by Mediterranean standards, has a pronounced east–west nutrient gradient, with much of the region remaining at or near oligotrophic levels (Harding et al. 1999).

Strong vertical stratification, reflected in a well-developed pycnocline within the water column, appears to be the sole proximate cause for nutrient-rich water from the Po River to spread east to the Istrian coast (Degobbis 1989). Anomalously low-density water over a well-developed pycnocline can occur in the northern Adriatic as a result of very high surface temperatures, low-salinity surficial water because of dilution from high spring discharge from the Po (Degobbis et al. 1979), or a combination of the two. Also, spread of Po-influenced water far to the east can be caused by delayed arrival of autumn wind storms, allowing a more extended existence of the pycnocline (Faganeli et al. 1985).

Studies of nutrient availability and trophic levels in the northern Adriatic typically recognize that the area should be divided into regions based on criteria such as annual productivity and patterns of seasonal fluctuations. Gilmartin and Revelante (1983) recognized four regions (fig. 5.5): a western eutrophic region highly influenced by inflow of nutrient-rich water from the Po; a northwestern mesotrophic region offshore of Venice that is strongly influenced by nutrients flushed from the coastal lagoons by spring tides; a less variable northeastern mesotrophic region possibly resulting from anthropogenic nutrients from the urban/industrial complex centered on Trieste (several relatively small Italian rivers flowing into the region almost certainly have an effect as well); and an eastern oligotrophic region most commonly influenced by low-nutrient water moving northward due to the Adriatic-wide cyclonic flow.

South of the northernmost coastal region, the overall year-round pattern is an easterly decrease in nutrients across the northern Adriatic, with nutrient levels along the Croatian coast closer to those of the more southerly parts of the Adriatic than to the Po-enriched westerly waters (e.g., Revelante and Gilmartin 1976; Fonda Umani 1996; Degobbis et al. 2000). The standing crop varies enormously from season to season in the overall mass m^{-3} and across the east–west gradient (fig. 5.6). By far the highest chlorophyll a values are found in the western shallow northern Adriatic (western end of Po delta to Rovinj transects) during the winter, when windstorms and cold temperatures result in vertical instability of the water column and in western confinement of the nutrient-rich water from the Po and other Italian

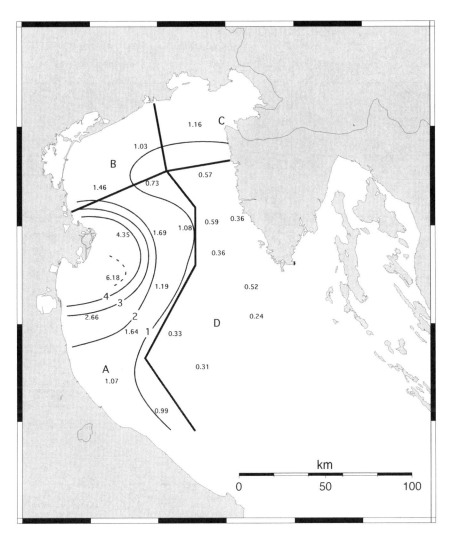

FIGURE 5.5 Mean northern Adriatic chlorophyll *a* standing crop (mg m⁻³) from 1972 to 1975.

From Gilmartin and Revelante 1983.

rivers. Western confinement of water from the Po River and also western concentration of the cold, high-density subsurface flow from the northern shallow region during the winter are clear in the more southerly Rimini to Pula transect.

The strong winter gradient between high northwestern and low southeastern chlorophyll *a* concentrations is diminished in the spring and autumn (fig. 5.6). A strong gradient forms again in the summer, but closer to the Po Delta, with oligotrophic conditions spreading throughout almost all of the northern Adriatic except for the area northwest of a line from approximately Venice to Rimini (lower left in fig. 5.6C).

chlorophyll a

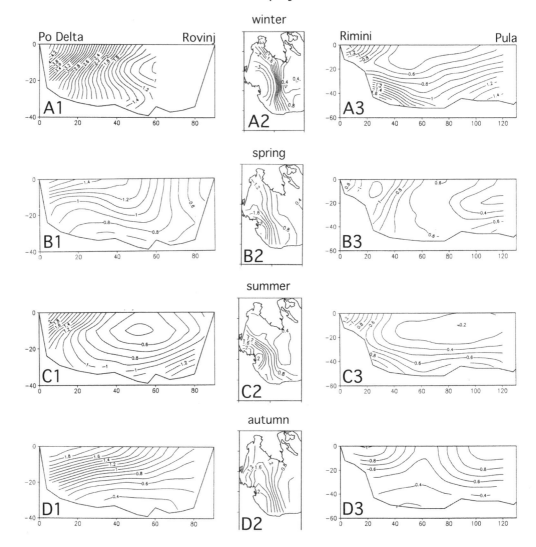

FIGURE 5.6 Chlorophyll *a* annual cycle in μg l⁻¹ for the northern Adriatic based on data collected from 1948 through 1991.

Reprinted from *Journal of Marine Systems*, vol. 18, Zavatarelli et al., "Climatological biogeochemical characteristics of the Adriatic Sea," pp. 227–263. Copyright © 1998, with permission from Elsevier.

Northern Adriatic Relative to Other Regions in the Mediterranean

The nutrient level of the northern Adriatic is on average higher than in other major regions of the Mediterranean Sea (e.g., Antoine and Morel 1995; Pettine et al. 1999; Zavatarelli et al. 2000). This is to be expected, inasmuch as the Po River is "the most important contributor of organic matter and nutrients to the Mediterranean" (Pettine et al. 1998:127; see also Zoppini et al. 1995; Pettine et al. 1999), and it is but one of several Italian rivers that empty into the northern Adriatic.

One of the most interesting aspects of the Adriatic Sea is the pronounced trophic gradient from high productivity in the northwest to typically oligotrophic, open-ocean conditions in the northeastern, middle, and southern regions, all of which are remarkably similar in nutrient levels and productivity. Overall, the more oligotrophic waters of the middle and southern Adriatic have chlorophyll *a* concentration less than half that of the north (Zavatarelli et al. 2000). The plume of nutrient-enriched riverine waters is closely confined along the Italian coast, maintaining a clear west to east gradient in productivity all the way to the Otranto Strait (Zavatarelli et al. 2000).

Phytoplankton

Planktonic bacteria and single-celled eukaryotes are sometimes reported by taxonomic group and at other times by size-class, the latter being a convenient way to sort the phytoplankton when conditions do not permit proportioning them into the taxa represented. Commonly used size classes are

picoplankton/ultraplankton	0.2–2 μm
nanoplankton	2–20 μm
microplankton	20–200 μm

Although nanoplankton in general have higher cell densities in the northern Adriatic (Revelante and Gilmartin 1983), the microplankton generally have higher standing crop biomass because the individual cells have an order of magnitude greater diameter and therefore about 10^3 times the volume of individual nanoplankton cells. However, nanoplankton have a shorter doubling time than do microplankton, and nanoplankton apparently account for more than half the primary production about 90 percent of the time in the northern Adriatic (Revelante and Gilmartin 1976;

Gilmartin and Revelante 1983). Diatoms are the most important group of microplankton in the nutrient-enriched northern Adriatic, while in the oligotrophic offshore waters of the southern and central Adriatic, dinoflagellates dominate (Fonda Umani 1996). The Adriatic therefore has the same nutrient-based distribution pattern of these two important groups of microplankton seen throughout the world oceans.

Overall abundance and taxonomic composition of the phytoplankton vary spatially and fluctuate throughout the year (figure 5.7) in response to temporal differences in the environment, many of which are predictable spatially and by season as described below. Extent of the various water masses that interact in the northern Adriatic is unpredictably variable. Interactions along the boundaries and mixtures of these water masses result in local and ephemeral vigorous population increases of various species, often resulting in a temporally changing patchwork of dominance. Nevertheless, throughout the Adriatic in general microplankton—primarily diatoms—bloom to high cell densities in late winter or early spring and in September–November. These blooms are in general related to winter overturn and to freshwater discharge maxima (Revelante and Gilmartin 1983).

Taxonomic composition of the microplankton in the northern Adriatic has been unstable for the past several decades, with large changes in the ratio of abundance of higher taxa (diatoms:dinoflagellates) and individual species within the higher taxa (Mioković 1999). Overall, the pinnate diatoms *Nitzschia* and *Pseudonitzschia* are the most abundant microplankton, although the dominant species within the two genera changed from the early 1970s to the mid-1990s. Diverse centrales diatoms also are abundant, e.g., *Skeletonoma*, *Cylindricus*, *Chaetoceros*, *Leptocylindricus*, and *Hagiosolen*. Overall, these centrales increased in abundance during the latter part of the twentieth century.

During the periods of conspicuous river input and presence of a prominent pycnocline, the cell density of microplankton in surface waters may be one hundred times that of bottom waters (Fonda Umani et al. 1992). In the Po-influenced surface waters and some other near-shore areas, microflagellates supplemented by centrales diatoms dominate during the summer, though locally in the summer-stratified water a mixture of dinoflagellates—such as *Protoperidinium* and *Scrippsiella*—and microflagellates prevails in surface waters (Fonda Umani et al. 1992; Fonda Umani 1996).

Heterotrophic bacterioplankton are involved in degradation of organic matter, and they also are primary producers utilizing both inorganic nutrients and DOM. Production by bacterioplankton is high in particular along the western coastal regions of the northern Adriatic where salinity

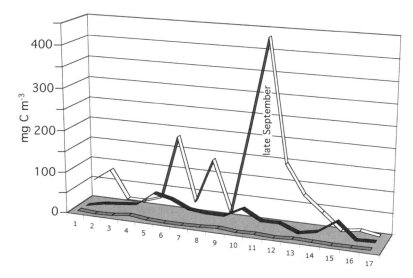

FIGURE 5.7   Pattern of biomass (mg C m⁻³) of net zooplankton (front, light gray ribbon) and the two main microphytoplankton fractions (dinoflagellates, intermediate dark ribbon; diatoms, white ribbon) for seventeen equal sample intervals from March 1986 through February 1987 in the Gulf of Trieste. Microzooplankton (not shown) have the same range of biomass as net zooplankton, and nanoplankton (not shown) have the same range of biomass as dinoflagellates.

After Cataletto et al. 1993.

is low and nutrients and DOM level are high. In this region they process 40–80 percent of the carbon synthesized by phytoplankton during productive periods. In winter their carbon demand can reach almost three times the relatively small amount of carbon produced by phytoplankton, which indicates that they use riverine inputs of carbon as well (Puddu et al. 1998).

SPATIAL PATTERNS OF PHYTOPLANKTON ABUNDANCE

It has been recognized for several decades that abundance of microplankton increases westward across the Adriatic (Revelante and Gilmartin 1976). This gradient in microplankton abundance parallels nutrient and chlorophyll *a* concentration as well as primary production rates, all of which reach their maximum within the eutrophic western coastal areas influenced by the Po River. Gotsis-Skretas et al. (2000) found that >5 μm diameter phytoplankton constituted about 80 percent of the chlorophyll *a* and of the primary production measures in the western area, whereas cells <5 μm were important in the oligotrophic eastern part of the northern Adriatic.

Both the <5 μm nanoplankton and the picoplankton had negative correlations with nutrient concentration, chlorophyll *a* concentration, and primary productivity.

Flagellates, most of which are nanoplankton, usually have higher cell counts (but not greater biomass) than diatoms even within the Po plume between Pesaro and Ancona, but the proportion of phytoflagellates is even higher at a distance of 30 km offshore, beyond the onshore Po plume (fig. 5.8). Diatoms, especially *Skeletonema costatum,* were dominant October 1990 through June 1991 in the onshore station but only for a short period offshore when the Po water was advected eastward (Zoppini et al. 1995).

Across the Po plume there is a high inverse correlation between biovolume measures and salinity (Zoppini et al. 1995; Giordani et al. 1997). Relationships between biovolume and individual nutrients introduced by the Po River are less clear. Giordani et al. (1997) found no significant correlation of biovolume with any specific nutrients, although Zoppini et al. (1995) determined that chlorophyll *a* across the Po plume correlates well with nitrate, silicate, and total dissolved phosphorus, although less well with orthophosphate. The higher correlation with total dissolved phosphorus indicates that in the low-nutrient waters of the Adriatic, organic recycling of phosphate is an important supplement to the external influx of inorganic phosphate (see fig. 5.4B; Ivančić and Degobbis 1987).

Photosynthetic efficiency increases with increasing nutrient concentration along the Italian coast of the Adriatic (Zoppini et al. 1995). This relates to the higher proportion of diatoms at high nutrient levels versus the higher proportion of phytoflagellates at lower nutrient concentrations. Phytoplankton require movement through water in order to acquire nutrients from water that they contact at the cell surface. Phytoflagellates are self-propelled and can therefore push themselves in any direction to encounter nutrients. Diatoms lack flagella or cilia and are entirely dependent on having a greater density than the surrounding water so that they can acquire nutrients from the water through which they sink. The higher the nutrient level of the water, the shorter the distance through which they must sink in order to double their mass and divide into two daughter cells. They are dependent on upwelling or wave-generated diffusion to lift at least half the daughter cells back up to the starting depth in order to maintain and increase population density.

The lateral advection of spring Po water induces an eastward shift of the larger sizes of suspended particles. In some years, such as those with a large spring flood of the Po River, low-salinity water is advected eastward across the sea to Istria and can mix through the entire water column of the eastern Adriatic (see figs. 4.9, 4.11). When this happens, the cell densities of microplankton (largely diatoms) near the east coast essentially equal those near the Italian coast, although delayed by about 3–4 weeks relative to maxi-

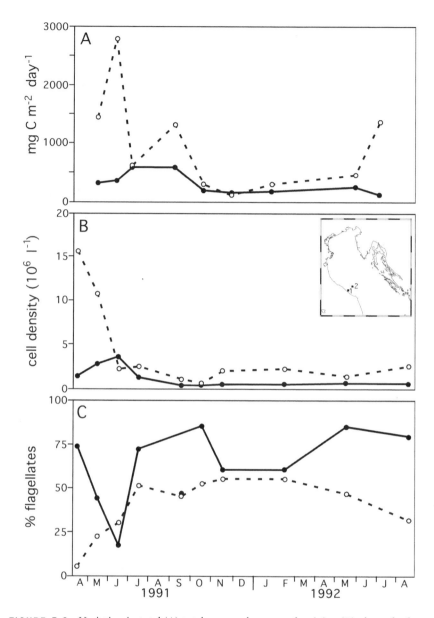

FIGURE 5.8 Variation in total (A) total water column productivity, (B) phytoplankton cell density, and (C) percent of phytoplankton constituted by diatoms for a seventeen-month interval in 1991 and 1992 at two stations offshore between Pesaro and Ancona, Italy (see inset map). Station 1 (open circles connected by dashed line) was within the Po plume at 11 m depth about 3 km offshore and station 2 (filled circles connected by solid line) was at 54 m depth about 28 km offshore, beyond the Po plume (except May and June 1991).

After data in Zoppini et al. 1995.

mum cell densities along the Italian coast (Revelante and Gilmartin 1992). Part of the increase in particle size is due to "microaggregates" of nano- and picoplankton (Revelante and Gilmartin 1992), which further changes the structure of planktonic food resources to grazers and suspension-feeders. Under more normal conditions there is a much higher proportion of small planktonic food particles available to suspension-feeders in the eastern part of the northern Adriatic than near the Italian coast.

Temporal variation in standing crop, productivity, and taxonomic composition of the phytoplankton in the northern Adriatic are clear in a one-year study of the phytoplankton offshore of the southeastern coast of the Gulf of Trieste (Malej et al. 1995). As for the northern Adriatic as a whole, the water column in the area studied is thermally stratified from May through September. The gulf is so shallow and contains such a small volume of water that the ratio of freshwater inflow (largely from the low-discharge Isonzo River) to total volume within the gulf is almost three times greater than for the northern Adriatic as a whole. Surface and bottom samples from sites >20 m deep yielded huge variations in phytoplankton cell counts and in relative abundance of various types of plankton through the year, apparently in response to several interacting environmental variables.

During the year of study (1992), the Isonzo discharge was lower than normal during the winter, but there were major floods in late March–April and again from October through early December (fig. 5.9A). The spring diatom bloom (fig. 5.9D) appears to have been in response to the introduction of nitrate by significant freshwater input during March and April (fig. 5.9B), before the water column was thermally stratified. The fresh water mixed first into the upper part of the water column, supporting a more vigorous surface bloom than occurred on the bottom. By May the spring bloom had crashed, because both the introduced nitrate and the remaining available silicate (fig. 5.9C) had been largely consumed by small- to medium-sized, rapidly proliferating diatoms such as *Skeletonema* and *Nitzschia*. Rainstorms in June, after the summer pycnocline had been established, introduced fresh water with an order of magnitude–higher N/Si ratio than that of the earlier river runoff (Malej et al. 1995). This produced a small bloom of flagellates (fig. 5.9D) which, in contrast with the diatoms, do not require silica for construction of skeletons. The flagellates were largely nanoplankton belonging to the Prymnesiophytes, secondarily microplankton consisting of dinoflagellates.

From April through September the water above the pycnocline maintained a constant, low concentration of silicate, while the silicate level in the bottom waters steadily increased to about nine times the original concentration (fig. 5.9C). The primary source of the increasing silicate was probably dissolution of diatom tests that had settled onto the sediment surface during the spring diatom bloom. The opaline silicates of planktic

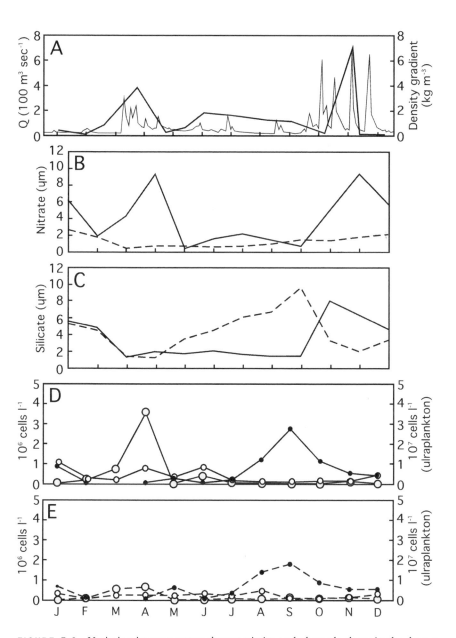

FIGURE 5.9 Variation in some water characteristics and phytoplankton in the deeper part (>20 m) of the Gulf of Trieste in 1992, by month. In B through E, solid lines represent surface water and dashed lines bottom water. In D and E, small solid circles represent picoplankton (<2 μm), small open circles represent nanoflagellates, and larger gray circles represent diatoms.

A) Rate of discharge (Q) of fresh water from the Isonzo River (light line) and difference in density between surface and bottom water in the Gulf (heavy line). B) Nitrate concentration in surface and bottom water. C) Silicate concentration in surface and bottom water. D) Plankton concentration in surface water. E) Plankton concentration in bottom water.

After Malej et al. 1995.

organisms are readily soluble in the northern Adriatic and do not last long as sedimentary particles. Only the most robust diatom frustules are abundantly preserved at greater than 10 cm depth along the margin of the rapidly accumulating Po-derived sediments, and the most delicate diatom frustules dissolve before they are buried (Puškarić et al. 1990).

Despite the increasing silicate concentration in bottom waters during the summer, Prymnesiophytes and green algae, rather than diatoms, dominated the phytoplankon at that depth (Malej et al. 1995). This dominance of swimming phytoplankton apparently was due to their ability to proliferate faster in low-nutrient waters than can diatoms, even in the presence of sufficient silicate for the construction of diatom skeletons. Picoplankton increased through the summer and reached their peak in both surface and bottom waters in September (fig. 5.9D, E).

Reduced insolation along with increased frequency and intensity of windstorms (table 4.1) changes the structure of the northern Adriatic water column in autumn from strongly stratified in September to vertically unstable and fully mixed from surface to bottom in December (fig. 4.16A, D). In 1992 this coincided with several intervals of high discharge of the Isonzo River (fig. 5.9A), which sequentially introduced low-density, nitrate-rich waters to the top of the water column. However, by the end of the year, vertical mixing of the waters was causing convergence in concentration of both nitrate and silicate in surface and bottom waters (fig. 5.9B, C). Thus, the condition of the water column became similar to that at the beginning of 1992, and with the spring increase in insolation and addition of nitrate-rich spring floods, the annual cycle of phytoplankton growth and nutrient fluctuation was set to begin again.

Phytoplankton are the primary source of nutrients for the benthic ecosystem, either through suspension feeding of benthos on phytoplankton or on microplanktonic grazers, or through settling onto the seafloor of the fecal pellets of grazers in the water column. It is obvious from the example above that there is seasonal variation in the abundance, size, and type of phytoplankton that are available, which—along with changes in temperature and other aspects of the physical environment—should yield seasonal variations in abundance, activity, or growth rate of all direct and indirect consumers, both within the water column and on the seafloor. However, there are conditions that can cause the normally operating annual cycle to cross a threshold and become toxic.

Algal Blooms I: Red Tides

Red tides are algal blooms that generate a reddish discoloration of the water. The red hue is due to a high concentration of dinoflagellates, which are a phylum (Dinophyta) of usually unicellular, golden to reddish-brown organisms characterized by a single transverse plus a single longitudinal flagellum. Dinoflagellates generally are photosynthetic but may be phagotrophs or live on decaying organic matter; some produce neurotoxins. Red tides occur most frequently in local coastal areas such as polluted harbors and less frequently extend across larger areas.

Red tides in the northern Adriatic are common in some coastal areas and as elsewhere in the world are most commonly due to blooms of single dinoflagellate species (Fonda Umani 1985, 1996; Fonda Umani et al. 1992). For example, successive dinoflagellate blooms occur locally along the Emilia-Romagna coast within the Po plume, due to high nutrient loads, inefficient grazing, rapid warming (in the spring), and freshwater inputs (Sellner and Fonda Umani 1999). In the late 1970s different spring through autumn blooms variously involved species of the dinoflagellate genera *Ceratium*, *Gymnodinium*, *Peridinium*, *Noctiluca*, and *Protocentrum* (Chiaudani et al. 1980; Sellner and Fonda Umani 1999). Perhaps a common trigger for these blooms was phosphate enrichment of coastal waters due to domestic and agricultural wastes from Emilia-Romagna, which supplemented the southward flow of the coast-hugging Po plume (Fonda Umani 1996), although one study (Penna et al. 2004) found that in autumn nutrients introduced from the Po were almost two orders of magnitude greater than the local sources.

ALGAL BLOOMS II: THE MUCILAGE PHENOMENON (MARE SPORCO OR DVJETNJE MORA)

Mare sporco ("dirty sea") and *dvjetnje mora* ("sea bloom") are the Italian and Croatian popular terms, respectively, for mucilage proliferation events in the sea. While small mucilaginous masses, known as "marine snow," are ubiquitous phenomena following marine algal blooms, the large-scale "mucilage events" in the northern Adriatic are a different matter altogether, characterized by huge blobs, stringers, and layers of the stuff. These masses form mainly in the upper water column and vary from a few millimeters to many meters across within the water column, in some cases producing a creamy to gelatinous surface on the sea (Stachowitsch et al. 1990; Precali et al. 2005).

These mucilaginous masses are generated by polysaccharide-rich exudates from diatoms that entrap suspended organic and inorganic material (Degobbis et al. 1999). In environments in which nutrient ratios approximate the Redfield ratio, high-molecular-weight polysaccharides constitute a higher proportion of exudates from stable populations of aged diatoms than when the populations are in exponential growth stages and consist of high proportions of young cells (Malej and Harris 1993). Although diatoms and their exudates constitute the bulk of mucilaginous masses, heterotrophic bacteria make up a substantial proportion of the masses and are important in the release of dissolved organic carbon back into the ambient water (Herndl and Peduzzi 1988).

The mucilage phenomenon has occurred at intervals at least back to the early eighteenth century (Fonda Umani et al. 1989). During the last two decades mucilaginous aggregates spread through the northern Adriatic during the summers of 1988, 1989, 1991, 1997, and 2000–2004. At its peak an intense mucilage phenomenon is characterized by flocks, strings, and clouds that eventually may be blown onto a coast or may remain offshore, where they settle onto a pycnocline within the water. During some mucilage events, extensive masses may form in open waters as turbid layers up to 0.5 m thick, composed of dispersed gel-like material that accumulates on a strong pycnocline between Po-diluted surface water lying over middle Adriatic–derived more saline water (Degobbis et al. 1995; Precali et al. 2005). The mucilaginous material, whether coalesced in situ or sedimented onto the pycnocline, settles to the seafloor later in the summer as the water column begins to mix following the summer stable period.

Mucilage events appear to be preceded by anomalously high concentrations of particulate organic carbon in surface waters influenced by riverine runoff (Ahel et al. 2005; Giani et al. 2005a). Production of mucilage corresponds with diatom blooms, although a rapidly decomposing dinoflagellate has been implicated as the possible primary producer of the mucilage in which diatoms are then lodged (Pompei et al. 2003). From the late 1980s different diatoms have dominated during the late spring than had been characteristic during the 1960s and 1970s (Degobbis et al. 1995). During the formation of late spring mucilage events, the same diatom species that occur within the mucilaginous aggregates are found in the surrounding water column, although older, larger mucilaginous aggregates are dominated by a different diatom community and lighter-weight fatty acids that slow the sinking rate of the aggregates (Najdek et al. 2002).

Mucilage events have followed abrupt increase—although not necessarily high discharge—in spring runoff (Cozzi et al. 2004), and pulses of nutrient-rich fresh water generate a series of blooms (Degobbis et al. 1999). Phosphorus is the limiting nutrient in this period, and its uptake by phytoplankton is very fast. With a strongly developed stratification of the upper water column, characterized by various pycnoclines due mainly to the pulses of fresh water, as well as intense heating of the sea surface, phosphorus can become unusually depleted from surface waters and thereby trigger high rates of release of polysaccharide mucus by phytoplankton (Urbani et al. 2005). At the same time, phosphorous limitation may significantly decrease degradation of the exudates by bacteria.

Mucilaginous aggregates support high densities of heterotrophic micro-organisms, predominantly heterotrophic bacteria but also choanoflagellates and monads (Herndl 1988). Within a vertically unstable water column, Herndl (1988) found a five-fold increase in bacterial concentration within aggregates over ambient conditions and also that bacterial decomposition of the aggregates stimulates proliferation of nanoflagellates (especially choanoflagellates and monads) in the surrounding water.

Only a few copepods, largely harpacticoids, are capable of grazing on mucilaginous aggregates, and the copepods that are most abundant within the northern Adriatic water column do not graze on them (Bochdansky and Herndl 1992). The polysaccharide exudate apparently is a deterrent to copepod grazing because it increases viscosity, therefore increasing the already high energy expenditure required for copepod swimming, and possibly also by release of distasteful chemicals (Malej and Harris 1993). The aggregates can, however, be inhabited by large numbers of grazing polychaete larvae and juvenile turbellarians (Bochdansky and Herndl 1992).

In conclusion, it appears that mucous events are favored by an unusually high N/P ratio due to climatological conditions and disturbance of the plankton communities, such as increased mucous exudation, modified bacterial activity, and reduced zooplankton grazing. They appear not to be simply a result of increased eutrophication of the northern Adriatic (Degobbis et al. 1995; Fonda Umani 1996; Giani et al. 2005a).

BENTHIC HYPOXIA AND ANOXIA

The normal November to March instability of the water column results in sea surface to bottom mixing so that bottom oxygen saturation is usually near 100 percent (fig. 5.10), essentially identical with that at the sea surface. Thermal stratification, which begins to develop with the increase in insolation and decrease in windstorms in the spring and lasts through October or mid-November, results in progressive decrease in oxygen saturation of bottom water. The bottom-water oxygen reaches a minimum in the period from September to mid-November, generally earlier in the shallower western regions (fig. 5.10). The decreasing oxygen saturation is due to a combination of low photosynthetic rate in bottom waters through the summer (fig. 5.9E), decay of particles settling through the pycnocline, and respiration by and decay of benthic organisms.

Bottom oxygen saturation typically is much lower in western regions strongly influenced by the Po outflow than in the eastern part of the northern Adriatic (Smodlaka 1986), as can be seen in figure 5.10. Delay in arrival of the autumn windstorms, however, can push the decrease in bottom

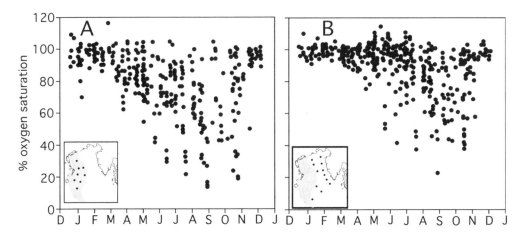

FIGURE 5.10 Bottom oxygen saturation annual cycle in the northern Adriatic based on data collected from 1966 through 1981 from the stations indicated in the inset maps. Data from the western stations most strongly influenced by water from the Po River are plotted in part A, and data from the eastern stations less frequently influenced by the Po River are plotted in part B.

After Smodlaka 1986.

oxygen saturation into November, which is reflected in the November-centered downward spike in the data fields in figure 5.10. Bottom oxygen can fall to such catastrophically low levels that benthic hypoxia and even complete anoxia can occur across large areas of the open northern Adriatic (for example in 1977 and 1989; Degobbis et al. 2000), as well as in some coastal regions, e.g., off Emilia-Romagna and in the Gulf of Trieste. Frequency and intensity of benthic anoxia is critically important in development and maintenance of benthic communities and is taken up at the end of chapter 7.

ZOOPLANKTON

Most zooplankton feed on phytoplankton or on other zooplankton and include animals that spend their entire life cycle in the water as well as the larvae or other planktic stages of otherwise benthic animals. As for phytoplankton, zooplankton are commonly grouped into size-classes rather than being specified by taxonomic groups. These are

nanozooplankton   2–20 μm
microzooplankton   20–200 μm
mesozooplankton   200 μm–2 cm
macrozooplankton   2–20 cm
megazooplankton   >20 cm

Nanozooplankton and some microzooplankton are actually single-celled het-
erotrophs such as radiolaria and foraminifera. Microzooplankton and mesozooplankton have the greatest biomass of the zooplankton size categories and
have been the focus of most zooplankton studies in the Adriatic Sea.

The Adriatic and adjacent Ionian seas have the highest species richness
and biomass of zooplankton per unit volume water in the eastern Mediterranean, although their values are more similar to those of the eastern Mediterranean and other oligotrophic areas than to the higher-nutrient waters
of the western Mediterranean (Kovalev et al. 1999). Maximum biomass of
microzooplankton (fig. 5.11) is found along the west coast of the northern
Adriatic, strongly developed during spring and summer when the water
column is stratified but minimal during the rest of the year. As seen in figure 5.11, the high abundance of zooplankton off the west coast is largely
due to high abundance within the surface, nutrient-rich, low-density water
associated with discharge from the Po and other Italian rivers. So, as one
would expect, high zooplankton biomass generally tracks high phytoplankton biomass and production (Benović et al. 1984; Revelante et al. 1985). Proliferation of zooplankton within the summer and autumn stratified water
column that is most influenced by the Po River discharge contrasts mark-
edly with the biovolume of zooplankton in the northeastern Adriatic and

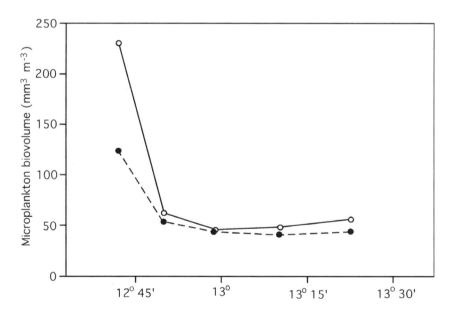

FIGURE 5.11   Mean 1978–1982 biovolume of microzooplankton in surface water (solid
line) and through the water column (*dashed line*) in a transect from off the Po Delta (*left
margin*) to offshore of the Istrian Peninsula (*right margin*).

After data in Revelante et al. 1985.

also in the northwestern Adriatic during the winter and spring (see biovolume given in fig. 5.12C–F). Average biovolume of zooplankton in the eastern north Adriatic (fig. 5.12D, F) is about 42 mm^3 m^{-3}, which equals biomass of 42–46 mg m^{-3} presuming near-neutral density, essentially equal to that in the middle and southern Adriatic (50 mg m^{-3}: Kovalev et al. 1999).

Copepod nauplii constitute the greatest biomass of the microzooplankton in both the eutrophic western and the oligotrophic eastern north Adriatic, and the only other microzooplankton group that is numerically and volumetrically abundant is the ciliates, including tintinnids (fig. 5.12). There are far greater numbers of ciliates throughout the year, but they are much smaller than are the copepod nauplii and postnauplii. Therefore, during the autumn and winter, when the water column is vertically unstable and mixed, they constitute a relatively small part of the microzooplankton biomass. However, the ciliates, especially the tintinnids, are sufficiently abundant during spring and summer when the water column is stratified, that total ciliate biomass exceeds that of the copepod nauplii and postnauplii in the west (fig. 5.12E; NTC plus TIN) and increases conspicuously even in the east (fig. 5.12F). Given sufficient nanoplankton food, tintinnids can proliferate sufficiently to form noticeable blooms in coastal regions (Fonda Umani et al. 1985). Microzooplankton-size meroplankton, which are planktic larvae of diverse benthic organisms, especially bivalves, are found in relatively low abundance (included in "Other" in fig. 5.12).

As would be predicted from the distribution of microzooplankton, which are the primary food source of mesozooplankton, the highest mesozooplankton biomass and abundance were found in the western eutrophic area influenced by discharge from the Po River, while the lowest values were found in the eastern oligotrophic area (Gotsis-Skretas et al. 2000). Throughout the Adriatic the most species-rich mesozooplankton group is the copepods, with chaetognaths, cladocerans, appendicularians, and meroplankton altogether making only a small percentage (Fonda Umani et al. 1992; Gotsis-Skretas et al. 2000). All of these except the cladocerans decrease in diversity from south to north. Within the northern Adriatic, nauplii and later-stage copepod larvae are numerically and volumetrically most abundant in nutrient- and microplankton-rich western waters, while adult copepods are more abundant in eastern, oligotrophic, nanoplankton- and picoplankton-rich waters (Gotsis-Skretas et al. 2000).

Fish

Although there is a fair diversity of pelagic fish in the northern Adriatic, few are sufficiently abundant to attract commercial fisheries. Sardines

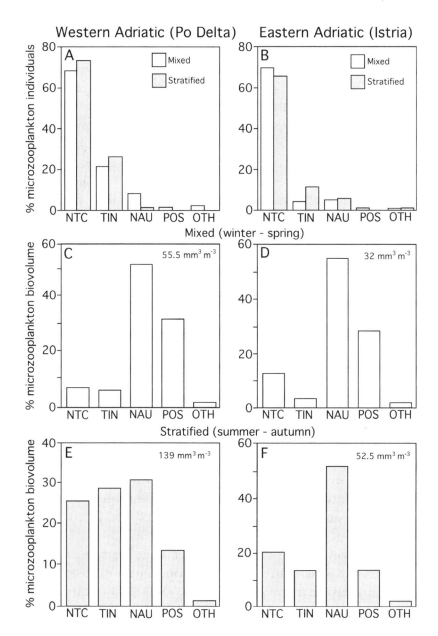

FIGURE 5.12 Relative abundance of the dominant groups of microzooplankton off the Po Delta compared with the region offshore of the Istrian peninsula. Data for the region off the Po Delta are given in A, C, and E, and data for the eastern Adriatic off the Istrian Peninsula are given in B, D, and F.

A, B) Percentage of the total number of individuals constituted by each group of microzooplankton. C–F) Percentage of the total biovolume of microzooplankton constituted by each taxonomic group during the season of mixed water column (C, D) and during the season of stratified water column (E, F).

Key to taxa: NAU = copepod nauplii; NTC = nontintinnid ciliates; OTH = all other taxonomic groups; POS = postnauplii copepods; TIN = tintinnid ciliates.

After data in Revelante et al. 1985.

(*Sardina pilchardus*), sprats (*Sprattus sprattus*), anchovies (*Engraulis encra-sicholus*), and mackerel (*Scomber scombrus*) have been most important. Bio-mass of pelagic fish across the northern Adriatic consists overwhelmingly of sardines, while the mackerel population crashed a few decades ago and the anchovy population declined during the 1980s (Bombace 1992; Houde et al. 1999). There are large yearly fluctuations in overall biomass of pelagic fish, which is a composite of strong fluctuation in biomass of individual species (Bombace 1992). Anchovies are more common along the Po-en-riched Italian coast, essentially tracking their zooplankton prey. This re-flects a positive correlation between anchovies, a planktic cladoceran, and the diatom *Nitzchia* (Stirn 1969 in Houde et al. 1999), presumably consti-tuting a direct food chain.

Overall, pelagic fish are not as directly important to the structure of benthic communities as are demersal fish. However, as high-level consum-ers within the water column they remove biomass from the water and transfer it to the seafloor as feces and carcasses.

SUMMARY AND IMPLICATIONS

Of the three primary nutrients, carbon is generally in excess in shallow wa-ter bodies, and availability of nitrogen or phosphorus is more likely to limit growth of primary producers and to have cascading effects through the en-tire ecosystem. Some skeletalized primary producers such as diatoms also require silicon in the same quantities as phosphorus. Availability of nutri-ents is determined by overall circulation pattern, temporal changes in cir-culation, locations and rates of external sources, and rate of nutrient recycling.

The two major external sources of nutrients into the northern Adriatic are the inflow of water from the middle Adriatic and the discharge of nutri-ent-laden fresh water from the Po and other rivers that enter the northern Adriatic between Trieste and Ravenna. Cyclonic flow from the middle Adri-atic introduces low-nutrient water into the northern Adriatic during winter; this water slowly sweeps northward along the Croatian coast and southward along the Italian coast. The winter cyclonic flow entrains the freshwater flow from the Po and associated rivers, confining it to a narrow band along the Italian coast and carrying it southward out of the northern Adriatic. During summer the decreasing density of the warming northern Adriatic water causes the cyclonic flow to fail, allowing the even lower-density fresh water entering from the Italian coast to pond and spread eastward.

As a consequence of the effects of the Po being largely confined to a near-west-coast band in the northern Adriatic, there is a strong west to east productivity gradient at all seasons. In general, the near-coastal western water varies from mesotrophic to eutrophic, even hypertrophic in some very restricted areas. Eastern water varies from oligotrophic to low-end mesotrophic. The intensity of nutrient and chlorophyll *a* concentrations varies seasonally and over shorter periods in both regions. Regardless, rate of production and standing crop of phytoplankton, microzooplankton, and mesozooplankton are always measurably and substantially higher near the west coast than in the eastern northern Adriatic.

The water column above any region of seafloor is the ultimate source of nutrients to animals on the seafloor, and the nutrients vary not only in quantity but also in quality and texture. Benthic suspension-feeders generally are specialized to consume particles of a particular size, some feeding predominantly on nanoplankton, others on progressively larger sizes. In addition, the seafloor is the recipient of materials falling out of the active ecological interactions that happen within the water column. Dead bodies of larger pelagic organisms may settle directly to the seafloor, but the rain of fecal pellets from zooplankton is more efficient in transferring useable nutrients from the water column directly to the seafloor, bypassing the benthic suspension-feeders.

The persistent contrast in productivity of western versus eastern water in the northern Adriatic should profoundly affect the ambient environment of benthos and the organic content of the sediments in the two areas. Plankton are more abundantly available to benthos in the west. Larger sizes of phytoplankton and zooplankton are more prevalent in the western than in the eastern northern Adriatic. More fecal pellets rain onto the sediment surface in the west than in the east, driving up the organic content of western sediments over that in the east. Higher organic content within the sediments and at the sediment–water interface in the west places a higher, persistent demand on the available oxygen. Lower productivity in the east means a smaller production rate of oxygen exists within the water column, which may be of little consequence in normal conditions. In the years of mucilage events, however, rapid delivery of mucous masses from the pycnocline to the seafloor can cause asphyxia of sedentary benthic organisms in the most mucus-choked regions, as described in chapter 7. Benthic anoxia caused by mucilage events is interesting and important, but for now at least it is a disruptive overprint on the profoundly different types of benthic associations that occur in different parts of the northern Adriatic.

CHAPTER SIX

PLEISTOCENE AND HOLOCENE SEDIMENTS

> We probably take a quite erroneous view, when we assume that sediment is
> being deposited over nearly the whole bed of the sea.
> —CHARLES DARWIN, *The Origin of Species*

With few exceptions, benthic communities across the northern Adriatic
live on and in an unconsolidated sediment floor. Particle size, sorting,
composition, source, and rates of deposition and resuspension are important aspects of the environment of benthic animals, because they determine physical and some chemical parameters of the seafloor on which the
animals depend. With the return of normal marine conditions following
the Late Miocene Messinian event, the Apennines, the Dinarides, and their
associated foredeeps continued their migration toward the axis of the Adriatic basin (chapter 3), and a persistent pattern of sedimentation was established that continues to the present.

The northern and much of the middle Adriatic Sea overlie a very low-
gradient, sediment-starved shelf with sediment accumulation predominantly along the west coast. In the central and eastern parts of the northern
Adriatic there are intermixed regions of erosion and of a thin accumulation
of Pleistocene and Holocene sediment. This post-Messinian pattern of sedimentation began to interact with sea-level fluctuation due to glaciation
that began in the Late Pliocene. As a result of glacially driven sea-level
changes during the Pleistocene, the position of marine-margin sedimentation has migrated back and forth from almost the western end of the Po
Plain to the middle Adriatic depression (MAD). The present patterns of
marine sediment accumulation and distribution of marine organisms are
part of a single time frame in the evolution of this cyclic system, which is

superimposed on the gradual progression toward complete infilling of the Adriatic basin by Apennine- and secondarily by Alp- and Dinarides-derived sediments.

The rate of influx of allochthonous sediment, its texture, and its interaction with currents and wave motion are as critically important to the development and function of benthic communities as are nutrient input and availability. This chapter gives an overview of the sources, textures, and rates of influx of sediments in the Adriatic, and the pattern of distribution of sediments across the Adriatic seafloor.

Controls Over Sedimentary Patterns and Sequences

Transgression of the Adriatic during Early Pliocene followed the Messinian (latest Miocene) drawdown of the Mediterranean and its satellite basins. The basinwide spread of the sea is particularly well recorded along the Italian coast and into the Po Plain. Along the entire western margin of the Adriatic, fine-grained siliciclastic sediments flushed off the Apennines have dominated over the basin-derived calcareous fraction in marine sediments since the earliest Pliocene (e.g., Crescenti 1971; Moruzzi and Follador 1973). Fan-deltas along the edge of the Apennines were the source of siliciclastic-rich turbidites that spread basinward into the Apennine foredeep along the middle and northern Adriatic, reducing maximum depth in the foredeep from bathyal to neritic (Ori et al. 1986).

Pliocene–Holocene sediments flushed into the Adriatic basin (including the Po Plain) were derived largely from the Apennines, have been trapped in the Apennine foredeep, and locally exceed 6,000 m thickness (fig. 6.1). The Apennine-derived sediments consist predominantly of clay- through sand-sized siliciclastics that are size-graded according to energy regime in fluvial through outer shelf environments; some sediment makes its way into the middle and southern Adriatic basins as turbidites (de Alteriis and Aiello 1993). The northwest end of the Adriatic basin has received sediment at a high rate, and the excess has been flushed down the axis of the basin by way of the Po River and its antecedents, filling the Adriatic basin from its western end toward its southeastern end.

Near the end of the Pliocene, stacked transgressive-regressive sequences began to accumulate, extending from the western end of the Po Plain to the edge of the MAD—as well as along the shallow margins of the southern Adriatic and elsewhere—due largely to sea-level fluctuation (Aiello et al. 1995; Monegatti et al. 1997) on an overall subsiding shelf (Ridente and Trincardi 2002). At about the same time, various tropical mollusks dropped out of the sequence, suggesting that the shallow Adriatic cooled with the

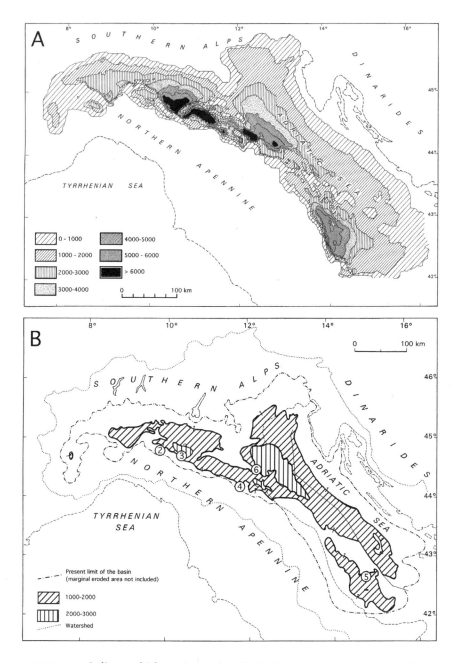

FIGURE 6.1 Sediment thickness (meters) in the Po Plain and northern and middle regions of the Adriatic Sea.

A) Pliocene to Holocene. B) Quaternary, depth below sea level to the base of Quaternary as defined on the base of the *Hyalinea balthica* zone. Although unshaded, the area between the present limit of the basin and the margin of the 1,000–2,000 pattern represents 0–1,000 m depth to the base of Quaternary sediments. Circled numbers are plotted at the localities for figs. 6.2 through 6.6; circled number 4 represents both fig. 6.3B and fig. 6.4.

Modified from Bartolini et al. 1996; reprinted with the permission of Cambridge University Press.

Late Pliocene onset of glaciation that drove the sea-level fluctuation (Mon-egatti et al. 1997).

PLEISTOCENE SEDIMENTS AND ENVIRONMENTS

Sediments exposed along the northern and northeastern margin of the Apennines comprise several sequences from the Early Pleistocene to the Holocene. Overall, the sequences indicate an excess of accumulation over subsidence of the Apennine foredeep in the Po Plain so that successive transgressions on average do not reach as far west as earlier ones had done.

Siliciclastic marine sediments were deposited as far west as the vicinity of Piacenza (fig. 2.5), at the western end of the Adriatic Sea, during at least one of the Early Pleistocene sea-level high-stands. A stretch of these sediments southeast of Piacenza, studied by Taviani et al. (1998), exposes a single shallowing-upward sequence, beginning with mollusc-rich sands deposited in an estimated 20–40 m water depth and grading upward to sandy shoreface sands (fig. 6.2). The sequence is unconformably overlain by Holocene alluvial sediments.

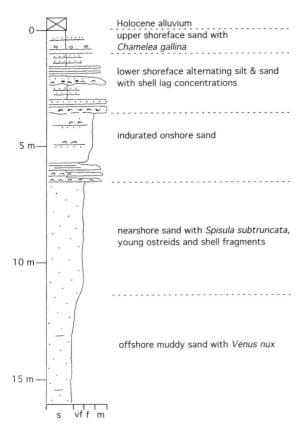

Holocene alluvium

upper shoreface sand with *Chamelea gallina*

lower shoreface alternating silt & sand with shell lag concentrations

indurated onshore sand

nearshore sand with *Spisula subtruncata*, young ostreids and shell fragments

offshore muddy sand with *Venus nux*

FIGURE 6.2 Columnar stratigraphic section of Lower Pleistocene sediments exposed along the Charo River southeast of Piacenza, Italy. See no. 2 on fig. 6.1 for geographic placement of section.

From Taviani et al. 1998. Courtesy of Società Paleontologica Italiana.

A more extensive Early Pleistocene marine record exists southeastward along the Apennine–Po Plain boundary. Northwest of Parma (fig. 2.5), Early Pleistocene sediments constitute an overall prograding sequence beginning with offshore sands overlain successively by shallowing-upward delta-front sands, then shore-zone sands and gravels, with some local lagoonal deposits at the top of the section (Di Dio et al. 1997). This sedimentary package, known as the Quaternario Marino supersystem, is unconformably overlain by later Pleistocene (Emilian) sands with gravel lenses increasing toward the top, deposited largely on an alluvial plain, and the Emilian system is overlain by three cycles of gravels grading upward to finer-grained sediments that represent alluvial fans developed within the past 470 Kyr (fig. 6.3A).

Up to seven stacked, fining-upward sedimentary sequences comprise the lower (Qm_1) part of the Quaternario Marino supersystem. Each sedimentary sequence ranges from 5 to 20 m thick and is characterized by a basal complex of sandstones, followed in succession by a bioturbated calcareous silty sandstone, a locally present shelly interval, and an upper mudstone (Dominici 2001). The high-diversity marine mollusk assemblages in the lower part of each sequence indicate shallow near-shore water, and the upper mudstones are generally characterized by a more offshore, lower kinetic energy *Venus* assemblage. Eastward pinchouts of the sandstones, the distribution of carbon-rich layers and other sedimentary features within the sequences, plus changes in nutrient levels suggested by the succession of

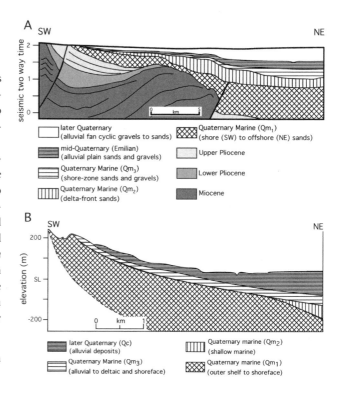

FIGURE 6.3 Two sections across the southwestern edge of Pleistocene sediments deposited in the Po Plain. See no. 3 on fig. 6.1 for geographic placement of sections. A) Stratigraphic and structural interpretation of AGIP seismic line from the edge of the Apennines to the edge of the Po Plain in the vicinity of Parma, Italy, with generalized depositional environments noted (after Di Dio et al. 1997). B) (*no. 4 on fig. 6.1*) Geologic cross section from the edge of the Apennines to the edge of the Po Plain midway between Bologna and Ravenna, Italy (after Amorosi et al. 1998b).

Courtesy of Società Paleontologica Italiana.

assemblages within sequences, indicated to Dominici (2001) that the sequences were generated by climate-driven changes at intervals less than 100 Kyr, with catastrophic floods bringing in the sands at the base of each sequence, and mudstones in the upper part representing return to a normal, inner-shelf setting in climatic conditions similar to the present.

Farther east along the base of the Apennines, along the stretch from Modena—which is near Parma—to the Adriatic coastal town of Rimini, the Pleistocene sequence is similarly dominated by clastic sediments flushed off the advancing tectonic front of the Apennines (fig. 6.3B). However, eastward along the Apennine trend a greater proportion of the sequence is marine (Amorosi et al. 1998a, 1998b, 1999a, 1999b). The Early to earliest Middle Pleistocene Argille Azzurre is stratigraphically equivalent to the lower two parts of the Quaternario Marino supersystem seen near Parma. The Argille Azzurre is overall a shallowing-upward sequence ranging from outer shelf deposits in the lower half to diverse shallow marine, shoreface, and back-barrier deposits in the upper few meters (fig. 6.4).

Middle Pleistocene Imola Sands overlie the Argille Azzurre and record two immediately superimposed fourth-order sedimentary cycles (Amorosi et al. 1998b). The Imola Sands are bounded top and bottom by the unconformities that define the upper part (Qm_3) of the Quaternario Marino supersystem. The lower part of the Imola Sands consists of sand and gravel interpreted as delta-front and littoral deposits. An unconformity separates this first, abbreviated cycle sequence from the overlying more complete sequence. The upper cycle of the Imola Sands consists at its base of alluvial clays and silts, followed conformably by lagoonal and estuarine sands and silty clays, separated by a ravinement surface from overlying fine sand with sedimentary structures that indicate basal shoreface and near-shore environments. The

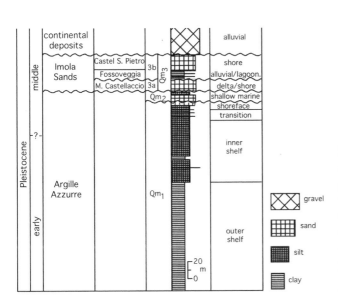

FIGURE 6.4 Composite columnar section of the Pleistocene stratigraphic sequence along the edge of the Apennines between Bologna and Ravenna, Italy, with generalized depositional environments noted. See no. 4. fig. 6.1 for geographic placement of section.

After Amorosi et al. 1998a, 1998b.

ravinement and overlying sediments represent rapid landward migration of a barrier and back-barrier system during a transgressive event. Fossils of brackish water-dwelling organisms occur at the very top of the unit, indicating an overall shallowing-upward trend during the sea-level high-stand. The two sequences within the Imola Sands are dated at approximately 800 Kyr and 700 Kyr ago, and represent two successive glacially driven marine transgressions at or near the Early to Middle Pleistocene boundary.

CYCLICITY OF PLEISTOCENE ADRIATIC SEDIMENTARY DEPOSITS

Stacked Middle and Late Pleistocene depositional sequences in the middle and northern Adriatic are interpreted to have formed at 100 Kyr intervals and represent multiple approximately 100–120 m sea-level fluctuations. Four Middle to Late Pleistocene sequences are present in the MAD (Trincardi and Correggiari 2000) and within the adjacent shelf deposits that constitute a several hundred kilometer–long narrow clinoform (Ridente and Trincardi 2005). The uppermost two of the sequences within the MAD are represented by successive pro-delta wedges separated by a mud drape (fig. 6.5). The pro-delta wedges were deposited during two low-stands, the upper of the two during minimum sea level of the Würm glacial interval, with the intervening mud drape between the two successive low-stands representing the intervening high-stand (Borsetti et al. 1995; Trincardi and Correggiari 2000). The upper of the two pro-delta wedges within the MAD presumably spans the end of the complete Late Pleistocene sequence into the Early Holocene.

Pro-delta wedges that lap over the edge of the MAD could potentially be derived from either the adjacent Balkan coast or the adjacent Italian coast. The carbonate-dominated Balkan drainage basins provide only a small amount of sediment, but an extended clinoform of coalesced deltas comprising bottomset muds and coarsening-upward foreset beds extended basinward from the middle Adriatic Italian coast during the Pleistocene (Ori et al. 1986). However, these local deltas were small and prograded at maximum rates of 0.019 m yr^{-1}. Offshore they are overlapped by a coeval set of southeast-dipping, muddier foreset beds that were the leading edge of the Po Delta, prograding at a minimum rate of 0.16 m yr^{-1} during Pleistocene low-stands (Ori et al. 1986). The Po Delta therefore was prograding at least ten times faster than the small, local Apennine-derived deltas, and it terminated in a depocenter 250 m thick (largely foreset beds) on the northern margin of the MAD (Trincardi et al. 1996).

North of the MAD there are at least three Middle to Late Pleistocene depositional cycles that underlie Holocene sediments and are due to

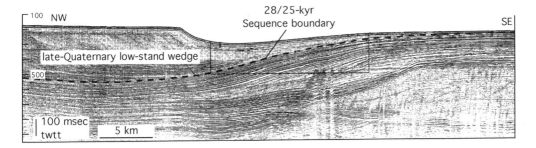

FIGURE 6.5 Seismic profile through a portion of the upper 400 Kyr sedimentary sequence across the northwestern edge of the middle Adriatic depression. In addition to the Late Quaternary low-stand pro-delta wedge, there is an earlier pro-delta wedge entering the section along the southeastern end; the two pro-delta wedges are separated by a mud drape that accumulated between the two successive sea level low-stands. See no. 5 on fig. 6.1 for geographic placement of profile.

glacially driven changes in sea level. Offshore of Ancona the cyclic sequences underlie the Holocene mud blanket swept southward from the present Po Delta, where each consists of alluvial plain sediments terminating in a pronounced erosional surface including sinuous channel cuts into the upper surface (Ferretti et al. 1986). The upper two cyclic sequences, plus the overlying Holocene sediments, are particularly well studied under the easternmost edge of the Po Plain south of the present-day Po Delta (Amorosi et al. 1999a, 1999b, 2003; Fiorini and Valani 2001).

Amorosi and colleagues closely sampled sediment in multiple cores penetrating up to 170 m of sediment and distributed over a 25 × 25 km area between Ravenna and the Po Delta. Below the Holocene sediments the cores cut through the complete Late Pleistocene cycle and into the top of the final Middle Pleistocene cycle (fig. 6.6). The region received terrigenous sediment at a high rate during the Middle and Late Pleistocene, delivered from the upslope length of the Po drainage basin. Each cycle begins at an unconformity between underlying alluvial plain sediments and overlying marine-margin sediments (back-barrier mud or shore-zone sand). The marine-margin sediments and some portion of the succeeding shallow marine muddy sands constitute a transgressive system tract. An otherwise ill-defined point within the shallow marine sediments that marks the attainment of highest sea level can apparently be determined by mollusk associations (Scarponi and Kowalewski 2004). This point is succeeded by shallowing-upward marine sediments with coastal sands on top, constituting the high-stand system tract.

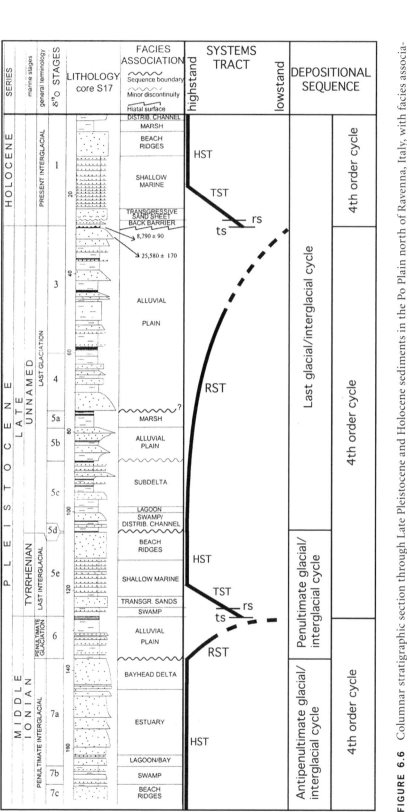

FIGURE 6.6 Columnar stratigraphic section through Late Pleistocene and Holocene sediments in the Po Plain north of Ravenna, Italy, with facies associations and their inferred environments, and relationship of the depositional sequences to glacial stages. See no. 6.fig. 6.1 for geographic placement of section.

Key: HST = high-stand system tract; RST = forced regressive system tract, TST = transgressive system tract; ts = transgressive surface; rs = ravinement surface.

Reprinted from *Quaternary Research*, vol. 52, Amorosi et al., "Glacio-eustatic control of continental-shallow marine cyclicity from late Quaternary deposits of the southeastern Po Plain, northern Italy," pp. 1–13. Copyright © 1999, with permission from Elsevier.

The high-stand system tract is succeeded by sediments deposited in environments transitional to an alluvial plain as sediment supply exceeded accommodation space and the shoreline—in this case the Po Delta—moved basinward, initiating a forced regressive system tract consisting largely of alluvial plain sediments. Each stack of alluvial plain sediments in the area studied was apparently initiated as the shoreline prograded while sea level was at or near maximum and likely continued to accumulate as sea level fell during the onset of the succeeding glaciation. However, by the time the sea reached its lowest level—approximately 120 m lower than at present—the edge of the Po Delta was about 250 km south, at the edge of the MAD (Late Quaternary low-stand wedge in fig. 6.5; Borsetti et al. 1995; Trincardi and Correggiari 2000; Asioli et al. 2001). Probably the studied area was entirely or largely bypassed by sediment during the low-stand so that little or none of the local regressive system tract was deposited during maximum glaciation.

Holocene Sediments of the Adriatic Sea

Fine sediments constitute almost the entire floor of the Adriatic Sea (fig. 6.7; Jenkins et al. 2005), except for local pebble beaches, rock outcrops, and basal coarse-debris aprons along and near the rocky shores, and locally lithified sands indicative of relict beaches or sediment outgassing. Sediment is introduced into the Adriatic around its entire margin, but by far the largest volume is derived from the Apennines, and the Po River is the primary point of input (Pigorini 1968; Frignani et al. 2005).

The Po introduces only fine-grained sediments into the Adriatic, 70 percent of which is silt, secondarily sand, and only 7 percent clay (Colantoni et al. 1979). Some of the sediment from the Po adds to the growth of its own delta. However, the majority of it is swept southward and accumulates along the Italian coast (Pigorini 1968; van Straaten 1971; Brambati et al. 1983; Boldrin et al. 1988; Faganeli et al. 1994; Cattaneo et al. 2003; Frignani et al. 2005).

Most of the sand introduced by the Po River is deposited in foreset beds on the delta front, at less than 10 m depth, and much of the silt settles at depths of about 15–25 m on the pro-delta surface, near the distributary mouths (Boldrin et al. 1988). However, especially during flood periods, the most intense of which occur during the winter, nonflocculated clays are carried in suspension for greater distances. During winter the western north Atlantic current is most vigorous, rapidly flowing, and narrow (fig. 4.8A), and the current rapidly moves the confined plume of suspended sediment southward along the Italian coast. The resulting belt of Po-derived mud can be traced over halfway down the Italian coast, at least as far south as the protected embayment immediately south of the Gargano Peninsula (Cattaneo et al. 2003; Trincardi et al. 2004; Frignani et al. 2005).

The ephemeral northern north Adriatic cyclonic gyre, which is north of a line from the Po Delta to the Istrian coast (fig. 4.4), is evident after bora windstorms (fig. 4.6) as a northeastward extension of low-salinity water drawn from the Po outflow almost to the northern Istrian coast (fig. 4.11). The turbid water traveling in this plume results in a muddy strip of sediment (fig. 6.7) that corresponds closely with the northeast-extending filament of low-salinity water following winter windstorms. However, only a very small percentage of the total of suspended matter travels northeastward, as can be seen in the thickness pattern of Holocene muds derived from the Po (fig. 6.8). At 20–25 m depth immediately in front of the main mouth of the Po River, Holocene muds are about 10–12 m thick, but they thin to only a few centimeters thickness by 8 km offshore; their maximum thickness of over 35 m is reached southward along the coast, off Ravenna and other points farther south (Boldrin et al. 1988; Correggiari et al. 2001; Cattaneo et al. 2004).

Sea-level high-stand was reached about 5,500 years ago, and since then the Po Delta has prograded as a lens-shaped accumulation of mud with the position of maximum accumulation migrating progressively seaward (Cattaneo and Trincardi 1999). Rapid flocculation of clays within the final several kilometers of the Po results in vigorous sedimentation immediately offshore of the mouth of the Po di Pila (Fox et al. 2004), where sediment accumulation rates currently exceed 6 g cm^{-2} yr^{-1} (Frignani et al. 2005).

As the sediment in the Po plume moves south along the Italian coast, it is supplemented by a smaller volume of sediment derived from the coast and smaller rivers and is separated into shore-parallel bands characterized by different grain sizes. Fine sand accumulates in a near-shore strip, paralleled by an offshore mud belt. Several sedimentary facies based on proportions of intermixed sand, silt, and clay have been defined in the western Adriatic north of the MAD (Poluzzi et al. 1985). The facies fall into three groups that appear to be related to three different kinetic regimes: shore sand and offshore relict sand belts that are—or were, if relic—most affected by wave motion; muddy sands in a narrow band just offshore, swept southward from the Po by the north and middle Adriatic currents; and broadly distributed shelf mud where clays have gradually settled beyond the influence of waves and currents.

Other zones of active sedimentation are located largely along narrow coastal strips. The location of the zones is determined by the interaction of sediment input and the prevailing current system in the Adriatic, as can be seen by comparing figures 4.4, 4.11, and 6.7. The current circulation pattern was established only about 6,000 years BP, when sea level rise

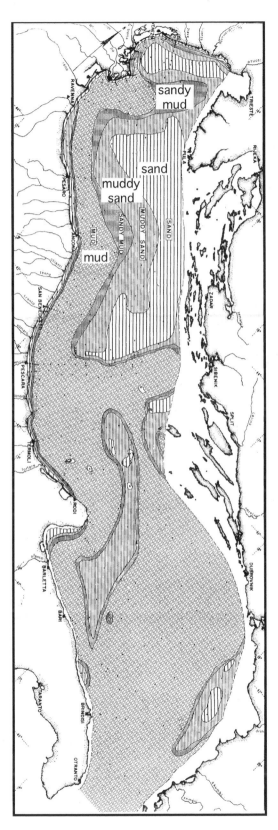

FIGURE 6.7 Sediment size-distribution across the floor of the Adriatic Sea.

Reprinted from *Marine Geology*, vol. 6, Pigorini, "Sources and dispersion of Recent sediments of the Adriatic Sea," pp. 187–229. Copyright © 1968, with permission from Elsevier.

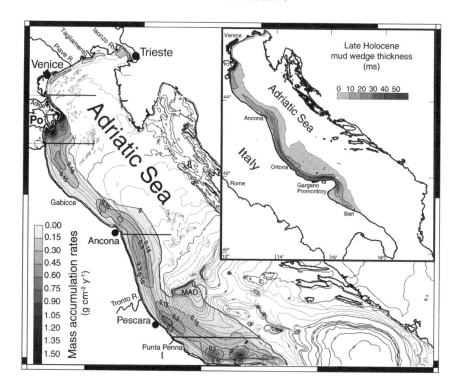

FIGURE 6.8 Past hundred-year mass-accumulation rate of sediment in the northern Adriatic Sea, based on ^{210}Pb, and (*inset*) thickness of the Holocene mud wedge contoured on sonar two-way travel time.

Reprinted from *Marine Geology*, vol. 222–223, Frignani et al., "Fine-sediment mass balance in the western Adriatic continental shelf over a century time scale," pp. 113–133. Copyright © 2005, with permission from Elsevier.

broadened the previously narrow connection between the MAD and the southern Adriatic as the entire width of the Pelagosa sill was flooded (Asioli et al. 1996). At the 6,000-year BP level in the continuous sedimentary record within the MAD, the planktonic foraminiferan *Globorotalia inflata* disappears and algal symbiont-bearing planktic foraminiferans increase (Asioli et al. 1996). This change is interpreted to indicate a rapid change from nutrient-rich to oligotrophic surface water as more Levantine low-nutrient water was introduced along with the change in the basinwide circulation.

All regions of active sedimentation other than the Po Delta and the plume-influenced area south of the delta receive relatively small amounts of sediments. Most sediment introduced south of the Gargano Peninsula is swept southward along the Italian coast, except some fines that migrate basinward. Sediment from Albanian rivers is swept northward along the Balkan coast of the southern and middle Adriatic, although much of it is lost via turbidity flows into the deep water of the south Adriatic basin (Pigorini 1968; Faganeli et al. 1994; Tomadin 2000). Some fine-grained sediment has continued to accumulate in the MAD during the Holocene, although sedimentation rate in the basin abruptly dropped 14,000 years ago to about 10 precent the previous rate of accumulation, at the time when sea level rose above the rim of the MAD and the region of deposition of sediment from the Po began to retreat from the MAD's margin (Asioli et al. 1996).

North of the entry of Albanian rivers into the Adriatic, only trivial amounts of terrigenous sediment are derived from the carbonates along the Dalmatian and Istrian coasts. Sediment introduced via several rivers that enter between the Po Delta and the city of Trieste at the northern tip of the Adriatic accumulates in a very narrow coastal, wave-influenced band where shore-parallel drift is variably toward the northeast or southwest but with a net movement toward the southwest (Pigorini 1968; Stefanon 1984; Brambati 1992; Tomadin 2000).

COMPOSITION OF ADRIATIC SEDIMENTS

Sediments accumulating in the Adriatic are largely fluvially introduced terrigenous and include primarily quartz, calcite, dolomite, aragonite, illite/muscovite, chlorite, kaolinite, Ca-montmorillonite, feldspars, and pyrite (fig. 6.9; Faganeli et al. 1994). Quartz—largely present in silt- and clay-sized fractions—is distributed rather uniformly, varying from 11 to 30 percent throughout the basin and averaging 15 percent. The highest clay content is in the southern Adriatic and the lowest is along the middle and northern Balkan coast. Overall, calcite is low along the Italian coast southward from the Po Delta and high along the Balkan middle and northern coast; it is dominantly detrital (Faganeli et al. 1994; Sondi et al. 1995).

In general within the middle and northern Adriatic, total carbonate content ($CaCO_3$ + $CaMg(CO_3)_2$) increases from the Apennine-influenced coast toward the Balkan coast, with the highest content off the Istrian coast and north of the influence of sediments from the Po River (fig. 6.10). Adriatic mud contains 20–50 percent carbonates (largely detrital; van Straaten 1971); many sands have an even higher carbonate content. Beach sands between Trieste and Venice average 80 percent carbonate, the percentage along local stretches being dependent in part on the different types of geologic terrains drained by each of the rivers that reach the coast between the Po Delta and Trieste (Venzo and Stefanini 1967). Between Trieste and Punta Salvatore (the northwest tip of Istria), the carbonate content diminishes toward the east because of the siliciclastic sediments entering via the

FIGURE 6.9 Mineralogical composition of sediments in the Adriatic Sea.

After data in Faganeli et al. 1994.

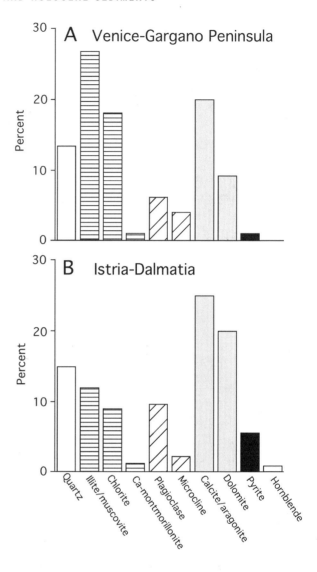

Rosandra River and eroding from the flysch that comprises the cliffs in the vicinity of Trieste.

All aragonite and a locally variable proportion of calcite in Adriatic sediments is biogenic (Faganeli et al. 1994). A higher proportion of calcite in beach sands on the Italian coast is biogenic than along the Balkan coast, and exposed relict beach sands may be relatively pure biogenic calcarenite, up to 95 percent carbonate. Diverse carbonate skeletons and skeletal debris comprise the sand grains, large benthic foraminiferans being overall more abundant than grains derived from mollusks, ostracodes, bryozoans, and echinoderms (Faganeli et al. 1994).

Organic carbon is highest (>1 percent by weight) in sediments of the northwestern Adriatic, along the Italian coast south of the Gargano Penin-

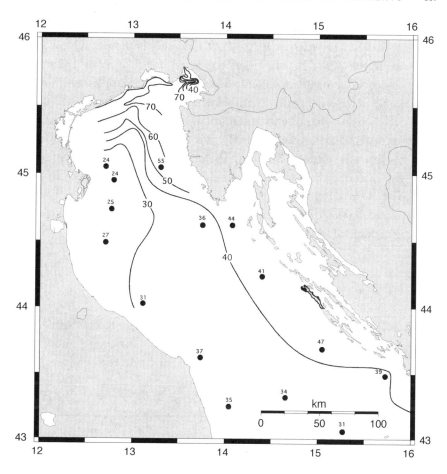

FIGURE 6.10 Percent total carbonate in surficial sediments in the northern and part of the middle Adriatic. Position of the 10 percent isograds north of 45°30' is from Venzo and Stefanini 1967 and is based on almost 200 sample locations; position of the isograds south of 45°30' is based on more widely spaced samples from Faganeli et al. 1994, which are plotted as numbered black dots on the map.

sula, and along the Albanian coast (Faganeli et al. 1994). The next-highest region (0.8–1 percent) is along the Italian coast from Rimini to the Gargano Peninsula, connecting the southern Italian region with the area adjoining the carbon-rich area off the north Italian coast. All these areas where the carbon content of sediments is high are adjacent to and downcoast from river mouths, and all have a $\partial^{13}C$ signature indicative of strong terrestrial input (Faganeli et al. 1994). Although the Po River is the ultimate source of high organic carbon content offshore of the Po Delta, the majority of sedimented carbon has an autochthonous rather than allochthonous fatty acid signature, having been processed through bacterioplankton and phytoplankton before accumulation in the sediment (Najdek and Degobbis 1997).

Authigenic pyrite occurs in the surficial 0–5 cm of sediments throughout the Adriatic (Faganeli et al. 1994). It normally is present within sediment as framboids ≤20 μm diameter or in foraminiferan chambers, indicative of reducing environment within sediments. Samples with weight of pyrite >3 percent generally occur below the 50 m isobath and are almost exclusively within a band of low total organic carbon content, low percentage of terrestrial carbon, and generally low sedimentation rates. The band stretches from the Otranto Strait, up the middle of the south Adriatic, and along the eastern half of the middle and northern Adriatic up the Dalmatian coast to the southern end of Istria. Lower concentrations of pyrite occur in coastal areas of high sedimentation rate, high carbon content, and high percentage of terrestrial carbon.

RELICT SEDIMENTS

Much of the northern and middle Adriatic receives little or no sediment at present (fig. 6.8), and relict Late Pleistocene alluvial sediments or more typically transgressive Holocene coastal sands constitute the seafloor (e.g.,

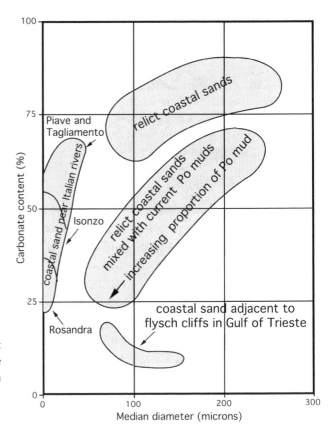

FIGURE 6.11 Carbonate content and grain size of surficial marine sediments found in the northern part of the northern Adriatic.

After Brambati and Venzo 1967.

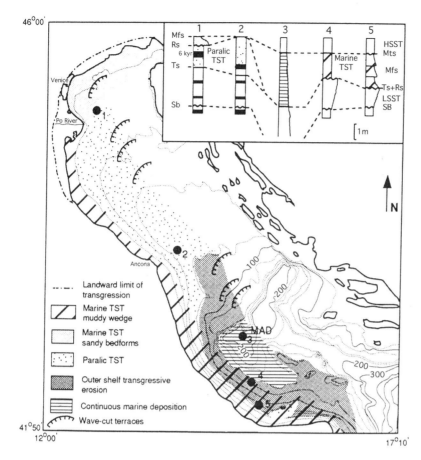

FIGURE 6.12 Transgressive deposits (18,000–6,000 BP) of the middle and northern Adriatic Sea, with notable wave-cut terraces indicated.

From Trincardi et al. 1994, © Springer-Verlag.

Brambati and Venzo 1967; Pigorini 1968; Brambati et al. 1983). The largest continuous area where relict sands are exposed extends from about the basin midline from just north of the MAD northward to the south side of the cross-Adriatic extension of the Po-derived muds (fig. 6.7). As indicated above, however, only a negligible amount of sediment is diverted into this cross-basin extension from the Po, and most of the sediment transported east is remobilized by storm waves and then carried by the cyclonic flow back into the coastal southward-moving system. Eastward of the well-developed western coastal zone of accumulation, residual Po-derived mud is only a veil that is mixed by bioturbation into the top of the relict sediments (fig. 6.11). The thin cross-Adriatic extension separates the larger southern area of clean relict sands from a somewhat smaller area of clean sands near the northwest coast.

Thickness of the transgressive sands is quite variable but averages only 30–40 cm (Colantoni et al. 1979). The Late Pleistocene through Holocene sea-level rise was rapid, averaging 10 m per 1,000 years from its position 120 m below present sea level 18,000 years BP to the maximum level 6,000 years BP. Sea-level rise was particularly rapid during the interval 11,000–6,000 years BP (Correggiari et al. 1996b) but more broadly tended to vary in rate on an approximately millennial scale (Amorosi et al. 2005). The interaction between the rapid but irregular Holocene rise in sea level and the very low gradient of the northern Adriatic floor resulted in a succession of back-stepping parasequences of sediment that do not completely overlap; consequently there is a submerged, progressively shallowing series of discrete shore-parallel barrier/lagoon system (paralic) deposits and wave-cut terraces (fig. 6.12; Trincardi et al. 1994; Correggiari et al. 1996b). Topography of some of the drowned barrier island and lagoon systems is remarkably well preserved (fig. 6.13; Correggiari et al. 1996a).

Narrow Eastern Belt of Sedimentation

There is some modern sedimentation along the near-shore eastern margin of the northern Adriatic (Fütterer and Paul 1976). In part this is due to erosion of small amounts of siliciclastic mud (terra rosa) from the adjacent carbonate landmass. In addition, skeletal remains of abundant benthic organisms accumulate in situ as whole skeletons, skeletal parts, and fragmented remains. Sand-sized and coarser skeletal carbonate is especially abundant within about 13 km of the coast, dropping precipitously (as does terra rosa–derived siliciclastic silt) offshore of this belt, where it is replaced by nonbiogenic relict dolomite and calcite sand.

A Local Sediment Trap: Cellaria Meadow off Istria

Lush growth of the erect, jointed bryozoan *Cellaria* at 35 m depth near the Istrian coast at Rovinj, Croatia, currently traps sediment as a biostrome of undetermined extent (McKinney and Jaklin 2001). Sea level did not rise to −35 m from the last low-stand until about 10,000 years ago (Fairbanks 1989), and remains of a coralline algal thicket underlie the *Cellaria* meadow at a depth of 15–25 cm within the sediment. (Bottom of the coralline-rich sediment was not reached by 25 cm–long hand-driven cores.) Coralline algal buildups in the Mediterranean may be initiated at about 10 m depth (Sartoretto et al. 1996), which suggests that sea level had reached −25 m or less before initiation of the bed of branched coralline algae below the *Cellaria* meadow, no more than about 9,000 years ago and probably more recently.

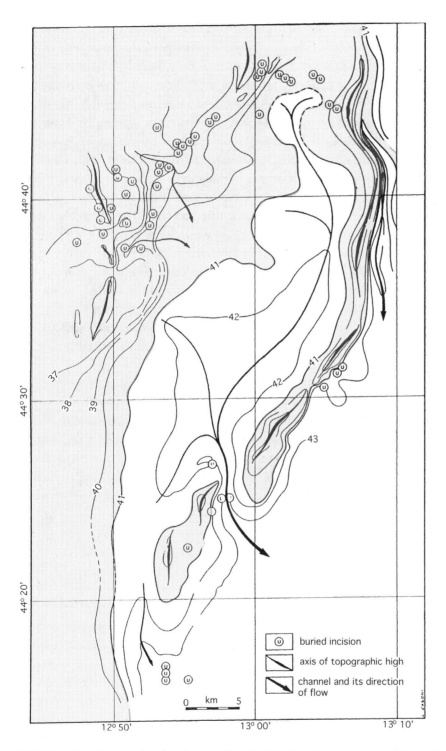

FIGURE 6.13 Topography of a submerged barrier island and lagoon system offshore of Ravenna, Italy. Contour lines are labeled in meters below sea level, and areas shallower than 41 m are indicated by gray shading.

From Colantoni et al. 1990.

Sediment above the coralline-rich sediment is uniform, with branch segments of *Cellaria* being the dominant megaskeletal component (fig. 6.14), and other skeletal remains within the sediment belong to taxa common in the living *Cellaria* meadow. The 15 cm of sediment overlying the coralline layer is interpreted to have been trapped by current-baffling of a continuous turflike growth of *Cellaria* (McKinney and Jaklin 2001) such as exists at the locality at present. The coralline-rich bed consists of over 50 percent silt-mud accumulated between the algal branches, and the *Cellaria*-rich sediment is 90–95 percent mud similar to that in the algal bed beneath. In both beds, slightly over 50 percent of the mud is insoluble.

All of the sand-sized and larger grains in the two beds are carbonate skeletons and skeletal fragments, almost all of which have a diameter >335 μm. These larger elements in the *Cellaria*-rich mud are mostly bryozoans and echinoderms (fig. 6.14B), while dominant large grains in the underlying coralline sediments are more diverse, including branched corallines, gastropods, bivalves, bryozoans, foraminiferans, and echinoderms. Echinoderm plates, especially those of ophiuroids and holothurians, foraminiferans, and sponge spicules, dominated the <355 μm sand fraction within the *Cellaria*-rich mud (fig. 6.14D).

The *Cellaria* meadow is developed near the base of an island around which tidal flow is focused over a belt of epibenthos-barren sand between the meadow and the island. A 25 cm core through the sand belt contained less than 50 percent mud, about 10 percent 65–335 μm sand, and more than 50 percent >335 μm grains throughout (McKinney and Jaklin 2001). The coarse grains in the sand belt (fig. 6.14C) were largely allochthonous but not derived from the *Cellaria* meadow: they consisted of about 45 percent bivalve and 30 percent gastropod shells and fragments. Finer grains within the sand belt also were more diverse than under the *Cellaria* meadow (fig. 6.14E).

SUMMARY

Surficial sediments of the Adriatic seafloor reflect largely the point sources of introduction of sediment along the coast, the overall circulation pattern, and the Late Pleistocene and Holocene patterns of deposition. The largest source of sediments for the Adriatic is the Po River, with important terrestrial sediment inputs also from rivers that enter along the Albanian coast and along the northern Italian coast between Trieste and Venice. In all instances the cyclonic flow within the Adriatic Sea transports the sediment along the coast, northward for Albanian rivers and westward and southward for the Italian rivers. The largest volume of sediment entering the Adriatic is silt, and only a small volume of sand and a trivial volume of coarser sediment is flushed in.

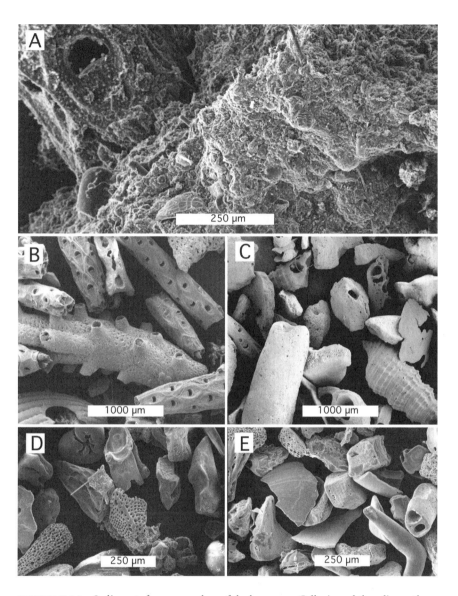

FIGURE 6.14 Sediments from a meadow of the bryozoan *Cellaria* and the adjacent bare sand offshore of Rovinj, Croatia.

A) Unprocessed sediment from 8–9 cm depth below the meadow; note segment of *Cellaria* branch visible in upper left. B) Sediment grains, largely *Cellaria* branch segments, greater than 335 μm from 3–4 cm depth within the meadow. C) Sediment grains, largely mollusk fragments, greater than 335 μm from 3–4 cm depth within the bare sand. D) Sediment grains, including ophiuroid plates and foraminiferans, from the 125–180 μm fraction within the meadow. E) Sediment grains, including ophiuroid plates and mollusk fragments, from the 125–180 μm fraction within the bare sand.

The highest rate of deposition is within the submerged part of the Po Delta and along the coast south of it. An ephemeral, far-northern cyclonic gyre carries a small stringer of mud eastward toward the Istrian Peninsula. The muddy stringer between the Po Delta and Istria interrupts a large area in which contemporary sedimentation is essentially absent and in which the sea is floored by a stepped series of Late Pleistocene and Holocene coastal sands deposited as the northern Adriatic was flooded.

Both carbonates and siliciclastic minerals are introduced via rivers. The carbonate content is highest in the eastern and northern portions of the northern Adriatic, in part because of the composition of the relict sands but also because of the accumulation of modern carbonate skeletons. Even near the Istrian Peninsula, within the region where sediment accumulation overall is slight or absent, local accumulations are developing where dense growth of erect organisms generate baffles that trap fine-grained sediments and provide local environments for growth of skeletalized animals and plants that add to the accumulating sediment.

Although sediment grain size is almost everywhere restricted to a narrow range of small sizes, there are sufficient local and regional differences in texture, composition, rate of deposition, and instability to affect recruitment and survival of benthos. The larger question to be addressed is whether the nature of the sediment transcends local effects to be the primary proximate cause of any larger, basinwide pattern of distribution of the benthos.

CHAPTER SEVEN

THE BENTHIC ECOSYSTEM

Bei Rovigno . . . schimmer zwischen den Algen meine Aumerksamkeit auf sich, und beim Auseinanderbiegen des Gebüsches fand ich dort—vie versteckte Ostereier—eine Reihe schönster Reteporenzoarien, die hier offenbar stets so wachsen, daß sie von den Pflanzen völlig überdeckt sind.

Near Rovinj . . . a glimmer among the algae caught my attention, and by spreading apart the tufts I found there—like hidden Easter eggs—a tier of the most beautiful reteporid bryozoan colonies obviously growing so well that they completely cover over the plants.
—HANS HASS (1948:22), pioneer in the use of SCUBA for scientific studies

After the introduction in chapter 1 of the general paleoecological question that this book is intended to address, almost nothing has been written specifically about the benthos of the northern Adriatic. The intervening chapters have been intended to give a basic overview of the geometry of the region and its dynamic interaction with eustatic changes in sea level, the geological development of the basin and the surrounding sources from which sediments are derived, the basic circulation pattern and the water masses involved, and the effect of that circulation pattern on nutrient and sediment distribution. All these topics are critically important for understanding the environmental context in which the benthic animals live. Now it is appropriate to turn to the animals themselves.

This chapter describes the benthos and its distributional pattern in the northern Adriatic. Description of specific taxa and their occurrence is addressed to some degree, but the primary focus is on the distribution and relative abundance of the constituent megaguilds of megascopic animals. (There is a substantial body of literature on foraminiferans and meiofauna of the northern Adriatic Sea, but although important these diminutive benthic organisms are not included here.) A summary of the regional patterns will be followed by closer examination of some of the more characteristic local benthic communities of the western and of the eastern coasts.

Benthos Across the Northern Adriatic

Aristocle Vatova came along at the right time. With the end of World War I, research in the northern Adriatic Sea was focused in only two surviving formerly Austrian marine laboratories, one in Trieste and the other in Rovinj. Vatova worked in the marine laboratory in Rovinj, and by the 1930s he was involved in a program to assess the benthos of the northern and middle Adriatic in order to understand better the food available to commercial fish stocks in the area. Between 1932 and 1934 he used a Van Veen grab with a 0.2 m² bite to sample approximately 150 sites offshore of Rovinj and during 1934–1936 expanded that by about 400 additional sites throughout the northern and middle Adriatic, from the Yugoslavian coast to the Italian coast as far south as the Gargano Peninsula (fig. 7.1).

Data provided by Vatova (1935, 1949) include the number of individuals and the biomass of all megascopic benthic animal species within each sample, which included both epibenthos and endobenthos. Agile epibenthos and some rare or delicate sedentary epibenthos are slightly underrepresented in grab samples (e.g., Ambrogi and Bedulli 1983), but grab samples better represent the relative abundance of epibenthos and endobenthos than do other methods of remote sampling. Vatova took only one—in

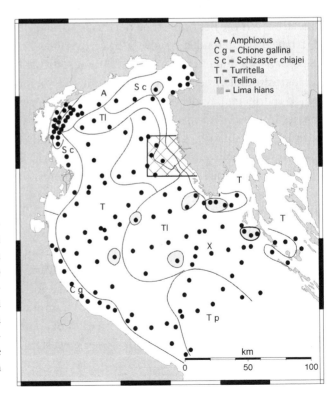

FIGURE 7.1 Locations of the northern half of stations at which Vatova (1949) sampled the benthos with a Van Veen grab in the late 1920s to mid 1930s. See fig. 7.9 for sample locations within cross-hatched area offshore of Istria reported by Vatova in 1935. The small X south of the Istrian Peninsula marks approximate location of specimens illustrated in fig. 7.17.

some instances two—grab sample per site, so local patchiness was not recognized. In addition, the inclusion of a single large-bodied animal can be quite conspicuous and may cause the single sample to misrepresent the local community to some unknown degree. However, the data are superb in that they cover the entire northern Adriatic, they were consistently collected by a reliable method that samples the entire local benthic fauna, and from the onset of World War II until the present there has been no set of comparable data covering the entire northern Adriatic. Two subsequent Italian grab sample–based surveys covered from the Italian coast east to the Italian-Yugoslav maritime boundary (Scardi et al. 2000). These two surveys therefore did not include the eastern third of the sea, where the benthos contrasts most strongly with that offshore of Italy! The two subsequent Italian surveys were made in the late 1960s and early 1990s and demonstrated the persistence of the species assemblage distributions recognized by Vatova.

Vatova (1935, 1949) recognized common cooccurrence among various taxa, which he identified as *zoocoenoses*. He produced a site distribution map in each of the publications to show the geographic distribution of each of the named zoocoenoses (figs. 7.1, 7.9). Gamulin-Brida (e.g., 1967, 1974) later chose to use the broader Mediterranean biocoenosis classification of Pérès and Picard (1964) in describing the benthos of the Adriatic. These two qualitative classification schemes do not organize information about the benthos in the way that I analyze it here, and they are referred to only sporadically in this and subsequent chapters. Instead, much of the analysis here goes directly back to the extraordinary set of raw data published by Vatova.

Taxonomic diversity of benthic animals is much higher in the northern Adriatic than the 328 species (plus 18 taxa or groups of taxa that could not be assigned to a particular species) recognized by Vatova. Nonetheless, with so many grab samples across the entire area, Vatova undoubtedly encountered most if not all the most widespread and abundant species, along with a sampling of the less abundant or more geographically restricted species. His data give a good idea of the overall taxonomic ratios within the northern Adriatic benthos. The most species-rich group identified to species in Vatova's samples was the Polychaeta, followed by Bivalvia, decapod Crustacea, Gastropoda, Holothuroidea, Ascidia, Ophiuroidea, and Sipunculida. Diverse other taxa were represented in Vatova's lists by five or fewer species each. (Species richness of higher taxa is given in more detail in chapter 8.)

Local standing crop biomass of the benthos varied more than two orders of magnitude across the northern Adriatic in Vatova's survey (fig. 7.2). One region of high biomass extended from the Po Delta northeastward toward the northern tip of the Istrian Peninsula. The other was a less well organized

FIGURE 7.2 Standing crop in g m⁻² wet benthic biomass across the northern Adriatic based on results of Vatova's sampling program.

Modified from Vatova 1949.

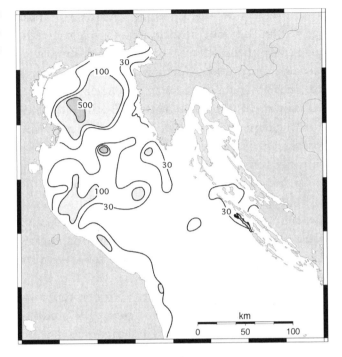

region that extended irregularly eastward from the vicinity of Rimini, Italy, with a narrow extension southward from Rimini along the coast. Benthic biomass peaked locally within these two areas at over 500 g m⁻², whereas at least half of the northern Adriatic, especially in the eastern portions, had standing benthic biomass of less than 30 g m⁻².

Within Vatova's samples biomass was highly skewed (skewness = 3.69; mode = 1 g/sample, median = 10 g/sample, mean = 35 g/sample). A natural break in mass occurred at 35 g of endobenthos per sample—or 185 g m⁻², since area of samples was 0.2 m⁻²—separating out approximately 12 percent of samples with highest biomass of endobenthos from the other samples. Interestingly, a natural break also occurred at 35 g of epibenthos per sample, and this break too separated out approximately 12 percent of the sites with highest biomass of epibenthos from the others.

Endobenthos (fig. 7.3A–B) was present in all samples except one nearshore sample off the Dalmatian coast (south of the Istrian Peninsula). It reached its highest biomass in the same regions as did total biomass, i.e., in an area extending northeastward from the Po Delta, and also in an area that extended irregularly offshore and southward along the Italian shore. High endobenthic biomass also occurred along the shore of the barrier islands of the lagoon of Venice, northeastward to the mouth of the Piave River, and in the Rovinj harbor. The endobenthos in these areas of high biomass was composed largely of polychaetes and bivalves, although an

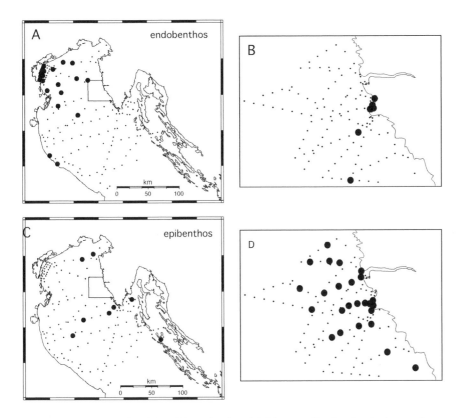

FIGURE 7.3 Maps of endobenthic and epibenthic biomass distribution in the northern Adriatic.

A) Endobenthos greater than 185 g m^{-2} indicated by large dots and lesser mass indicated by small dots with (B) detail of the densely sampled area off the central Istrian coast. C) Epibenthos greater than 185 g m^{-2} indicated by large dots and absent or lesser mass indicated by small dots with (D) detail of the densely sampled area off the central Istrian coast.

From data in Vatova 1935, 1949.

endobenthic suspension-feeding gastropod, *Turritella*, reached densities up to 1,200 m^{-2} and constituted almost all the biomass in some sites 25–50 km east of the Italian shore.

Epibenthos were absent from just over 20 percent of the samples, almost all of which were in the western regions. Epibenthos reached its highest species richness, number of individuals, and biomass (fig. 7.3C–D) along the Istrian coast and in isolated midbasin localities south of the Po Delta. The most widespread and abundant epibenthos were sponges, polychaetes, mollusks, holothurians, ophiuroids, and ascidians. Subsequent to Vatova, almost all biologists who have assessed the benthos across the northern Adriatic have commented on the low abundance of epibenthos in the west and its increase toward the east.

A notable aspect of the east–west contrast in benthic organisms is that the epibenthos characteristic of eastern communities appear to be more long-lived on average than the endobenthos characteristic of western communities (Ott 1992). This contrast appears to stand as a long-term difference related to there being greater environmental instability in the west than in the east. Increased frequency of anoxic events in the east during the latter part of the twentieth century profoundly disturbed communities in some regions, as described below. Several of the epibenthic organisms that characterize the communities off the Istrian coast typically have decade-scale life spans (or at least did before the increased frequency of anoxic events). Among these are the bryozoans *Cellaria fistulosa* and *Pentapora fascialis* (see below), and the bivalves *Arca noae*, *Pecten jacobaeus*, and *Pinna nobilis* (Zavodnik et al. 1991; Mattei and Pellizzato 1996; Peharda et al. 2002; Richardson et al. 2004).

Vatova's huge data set clearly demonstrates the east–west contrast in distribution of endobenthos and epibenthos. Much, much more can be gleaned from it, but more detailed use of his data is largely deferred to chapter 9. The current chapter is dedicated to introducing the range of ecological structure that can be seen in specific western endobenthos- and eastern epibenthos-dominated assemblages.

The basinwide patterns described above consisted of a mosaic of local assemblages that in general have persisted and have been found in more recent studies, except in local areas that have been most severely disrupted by eutrophication-driven events. The sections that follow summarize studies of a few typical and particularly informative assemblages from the western and from the eastern regions.

WESTERN BENTHIC ASSEMBLAGES

Western assemblages are overwhelmingly dominated by endobenthos, largely bivalve mollusks and polychaete annelids. They vary in taxonomic richness, in relative dominance of bivalves, and in whether the organisms feed largely within the sediment, on the sediment surface, or from just above the sediment–water interface.

PO DELTA NEAR-SHORE

The Po Delta extends seaward to a terminal point that projects about 20 km away from an otherwise roughly linear coastline. On the delta, the Po divides into several distributaries that would construct a birdsfoot pattern except that longshore currents redistribute sediments from what would be distally extending river banks into small barrier islands that extend across semi-enclosed bays (fig. 7.4). Human efforts to stabilize the delta in order to prevent its encroachment on the lagoon of Venice have also had a hand in shaping it (Stefani and Vincenzi 2005). Diverse, temporally fluctuating benthic communities are found in the bays and have been the focus of numerous ecological studies (e.g., Poluzzi and Forti 1983; Parisi et al. 1985; Bianchi and Morri 1996; Marchini et al. 2004). However, it is the offshore, open marine conditions along the delta that are the focus here.

In the shallow water within a few km of the delta (fig. 7.4: localities P), there is a low-diversity benthic fauna comprised largely of polychaetes and bivalves in fine to medium sand at 2.5 m depth to progressively muddier sand at 5 and 8 m–deep stations (Parisi et al. 1985; Ambrogi et al. 1990). Species richness varies from year to year at each depth but is higher at the deeper stations than at 2.5 m (Ambrogi et al. 2001). Many of the species are short-lived, and a large part of the variation from year to year is the result of highly uneven recruitment, which can vary more than four orders of magnitude even for the more common species (Ambrogi and Occhipinti Ambrogi 1985, 1987). Endobenthos constitute more than 90 percent of the fauna at all stations. Seasonally from 80 to 94 percent of the endobenthic individuals in the shallowest stations (2.5 m) are the suspension-feeding

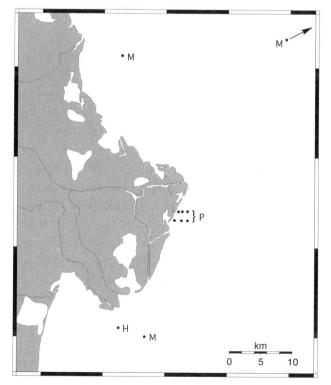

FIGURE 7.4 Po Delta with nearby benthic sampling stations described in the text.

H = Hammond et al. 1999; M = Mancinelli et al. 1998 and Simonini et al. 2004; P = Parisi et al. 1985.

Base map generated by Online Map Creation (www.geomar.de/omc/omc_intro .html).

bivalve *Lentidium mediterraneum*, which can occur in densities up to 300,000 m^{-2}. More diverse endobenthic suspension-feeding bivalves and polychaetes constitute from 12 to 72 percent of the total endobenthos in the deeper stations. The deeper stations have an increased proportion of deposit-feeding endobenthos. Sessile epibenthic individuals make up but 5 percent of the total at 2.5 m depth, essentially consisting entirely of barnacles growing on *L. mediterraneum* valves, and sessile epibenthos are almost completely absent at the deeper stations (Parisi et al. 1985).

Environmental conditions are more variable at the shallowest, nearshore stations off the Po Delta occupied by Parisi et al. (1985). Wave motion and consequent physical instability of the sediment floor, winter to summer temperature range, and salinity vary more at 2.5 m than at 5 and 8 m depths, and sites at all three depths are affected by strong interannual fluctuations in nutrients, oxygen concentration, and degree of admixture of outflow from the Po (Ambrogi et al. 1990). All stations have an annual pattern of cold-weather mass mortality followed by summer recruitment, which results in greater extremes in population density at 2.5 m than at deeper localities. By spring the population density at 2.5 m drops to about 6 percent of the summer level, largely because of intense mortality of *Lentidium mediterraneum*, whereas the dieback reaches only 25–50 percent at 5 and 8 m (Parisi et al. 1985).

Although the benthos offshore of the Po Delta is overwhelmingly endobenthic, some rapidly growing, environmentally tolerant sedentary epibenthos occur along the delta beaches where suitable substrata such as mollusk shells remain exposed sufficiently long. These include weedy species of widely distributed encrusting bryozoans (3), serpulid polychaetes (2), and barnacles (2), plus young bivalves (3) (Poluzzi and Agnoletto 1988). Altogether these species plus a soft turf of intergrown unidentified and soft clonal animals and plants covered only about half the available hard substrate space, with the other half remaining devoid of multicellular organisms.

Stations somewhat more distant from the Po Delta were sampled by multiple box cores by Mancinelli et al. (1998) and by Simonini et al. (2004), the most distant being about 50 km offshore (fig. 7.4: localities M). Nearshore and offshore sites were reported as strongly dominated by endobenthos, and no sedentary epibenthos were encountered at any of the stations. The two more inshore stations were in well-sorted silty clay, averaged very high in organic content (8–9 percent total dry weight of sediment), and were characterized by very high density and biomass of the endobenthic bivalve *Corbula gibba*. The more distant station was in poorly sorted silty sand beyond the strongest effects of the Po plume and had lower average organic content (3 percent) and greater species richness, was dominated by endobenthic surface deposit–feeding polychaetes and bivalves (but not

Corbula gibba), and was supplemented by carnivorous/omnivorous, suspension-feeding, and subsurface deposit–feeding polychaetes.

The offshore station of Mancinelli et al. (1998) and Simonini et al. (2004) had a more extensive vertical penetration of taxa than did the two near-shore stations. The upper 2 cm of sediment in the two near-shore stations were populated almost entirely by surface deposit–feeders, including *Corbula gibba*, which is a combination surface deposit–feeder and suspension-feeder. In the offshore station the upper 2 cm had surface deposit–feeders supplemented by carnivores/omnivores and some suspension-feeders. All stations showed a decreasing number of species in deeper zones within the sediment (2–5 cm, 5–15 cm), with deeper zones populated by carnivores/omnivores and by subsurface deposit–feeders.

Biomass was anomalously high 0–2 cm deep in the near-shore station south of the Po Delta but fell off abruptly at 2 cm depth to equal the other two stations in the 2–5 cm level. The greater abundance of opportunistic species such as *Corbula gibba* (see below), the concentration in biomass into surface deposit–feeders, and the slight penetration into deeper levels in the near-shore stations suggest that the high nutrient rain from the Po plume onto the sediment surface contributed to focusing benthic feeding on the sediment surface to the detriment of other trophic groups.

OFFSHORE OF RAVENNA

During the 1930s Vatova (1949) occupied a series of ten stations parallel to the shore between Ravenna and Pesaro, and several of the more open-water stations were clustered offshore of Ravenna. There was an overall change in species composition of the benthos from nearer shore, shallower, and sandier stations to further offshore, deeper, and muddier stations. All stations were dominated by endobenthos, with mobile epibenthos varying from absent to less than 15 percent of the biomass, and with sedentary suspension-feeding epibenthos almost completely absent (fig. 7.5). Diverse bivalves, gastropods, echinoderms, and polychaetes were present, and a few other higher taxa were represented by single species.

During summer 1985 multiple samples were collected at 14–15 m depth offshore of Ravenna (Crema et al. 1991) within the Ravenna to Pesaro region previously sampled by Vatova. About 50 percent greater surface area was included within their grab samples than Vatova had acquired in the vicinity, with densest sampling within a region of about 15 km diameter, and with interestingly different results from those of Vatova. Over 95 percent of the individuals and biomass were endobenthic, which falls within the range of values obtained by Vatova. Among the 19 most abundant taxa, 15 were polychaetes, 3 were bivalves, and 1 was an amphipod. However, a single

FIGURE 7.5 (A) Biomass and (B) number of individuals of benthos offshore of Ravenna during the mid-1930s, plus a bar along the right side of each indicating the benthos off Ravenna at 14–15 m deep in 1985. Station depths for the 1930s data grade almost monotonically from 8 m closest to shore to 40 m farthest offshore. *Corbula gibba* comprises over half of the biomass and number of individuals in Crema et al.'s data for 1985 but was uncommon in the 1930s.

Data from Vatova 1949 and Crema et al. 1991.

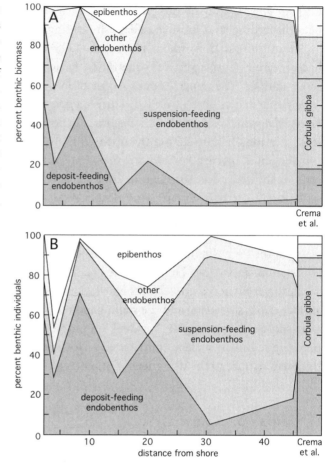

bivalve, *Corbula gibba*, averaged 51 percent of the individuals and 46 percent of the biomass at the stations, which was about five times the next-most abundant species by either measure. *C. gibba* had been present as a minor part in only one of Vatova's grab samples within the region, although a little to the north of Ravenna it was beginning to appear in very high numbers by the late 1970s (Hrs-Brenko 1981; see previous section).

Corbula gibba is a highly invasive species, a pioneer on defaunated bottoms, tolerant of unstable and hypoxic conditions, capable of anaerobic metabolism, and able to switch between suspension feeding and surface deposit feeding (Bonvicini Pagliai et al. 1985; Diaz and Rosenberg 1995; Pisarović et al. 2000; Simonini et al. 2004). The hyperabundance of *C. gibba*, along with the ubiquitous presence in Crema et al.'s samples of an instability-tolerant polychaete species, suggests that the frequent anoxic events along the Ravenna coast during the 1980s had reformulated the benthos in the region, without sufficient time since the most recent anoxic event to reestablish a stable community.

A further change in the shallow region offshore of Ravenna followed a major discharge of floodwaters from the Po in November 1996 that caused a benthic die-off, with the bioturbating amphipod *Ampelisca diadema* replacing *Corbula gibba* as the dominant species (Occhipinti-Ambrogi et al. 2005). Even though *A. diadema* had vigorously fluctuating population densities through the remainder of the study interval (July 1996–July 2002), *C. gibba* had only one brief interval of high abundance. During the interval, species richness gradually increased through two additional Po flood–instigated die-offs and mucilage events that had little or no effect on the benthic assemblage.

Although the benthos off Ravenna has been dramatically altered since the early twentieth century, the changes have been within a largely unchanged trophic system. The system has remained resolutely endobenthic right through the change in taxonomic dominance and community structure. Along the right side of figure 7.5, bars—shaded in the same way as the data from Vatova's (1949) offshore sequence—represent the proportion of megaguilds in the data of Crema et al. (1991). Despite the prevalence of *Corbula gibba* as the dominant part of the endobenthic suspension-feeders, the mix of endobenthic deposit- and suspension-feeders, endobenthic carnivores/omnivores, and minimal epibenthos is close to the average of Vatova's samples in the shallower waters within 20 km of the coast. Even the switch to dominance of *Ampelisca diadema* following the high 1996 Po discharge (Occhipinti-Ambrogi et al. 2005) remains within the same context of endobenthos feeding largely on or just below the sediment surface.

BIOTURBATION UNDER THE PO PLUME

Some of the endobenthic animals off the Italian coast are sedentary, extending deeper into the sediment only as they grow. These include many of the suspension-feeding polychaetes, and lateral movement of some of the suspension-feeding bivalves is stately in the extreme. Other bivalves, such as *Nucula* and *Tellina*, are deposit-feeders and therefore just like the deposit-feeding polychaetes must churn through the sediment in order to feed. All deposit-feeders, endobenthic carnivores, and epibenthic sediment-probing carnivores are vigorous bioturbators. In addition, some epibenthic crustaceans and demersal fish cover much of their bodies with sediment as camouflage. The activities of these bioturbators both reshape the structure of the sediments and alter benthic nutrient exchange rates.

All bioturbators open passageways through the sediment that to a greater or lesser degree connect with the sediment–water interface. They therefore provide large-diameter passageways through which water can flow into and out of the sediment. Biofilms flourish on the walls of open channels (e.g.,

Dworschak 2001) and accelerate the release of nutrients from the sediments into the water.

Rates of accumulation of muddy sediment directly offshore from the Po Delta are very high, about 1 g cm^{-2} yr^{-1} at sites about 5 km offshore, with low biomass of macrofauna and approximately 50 percent burial rates for carbon, nitrogen, and silicon (Hammond et al. 1999). Where sediment is most vigorously accumulating just off the mouth of the Po di Pila, it exceeds 6 g cm^{-2} yr^{-1} (Frignani et al. 2005). Fluxes of solutes from the sediment at these sites have rates similar to those calculated for diffusive flux.

Hammond et al. (1999) also studied flux of solutes in a series of stations between Ravenna and Ancona, extending through the Po plume from onshore silty sand, through sandy silt, and out into sand of the mid-Adriatic at 30 m depth. The sites under the Po plume in this transect had moderate accumulation rates of about 0.1–0.2 g cm^{-2} yr^{-1}, abundant macrofauna, and 5–20 percent burial rates for carbon, nitrogen, and silicon. Fluxes of solutes from the sediment at these sites are at approximately twice the rates calculated for diffusive flux. In the farthest offshore site, beyond the Po plume, there was virtually no sedimentation and macrofauna were abundant, but burial efficiency for nutrients was not determined.

There is a strong contrast in shallow sediment pore-water chemistry profiles between all of the stations in the southern transect and the near-Po stations. There is no discernible vertical pattern of concentration of pore-water solutes in the southern transect, while the near-Po stations have a stable concentration at depth with a rapid fall-off in concentration within the surficial 5 mm, approaching the value for ambient water (fig. 7.6; Hammond et al. 1991). At the southern sites, sediments are strongly irrigated by bioturbation, increasing the solute flux back into the water, whereas the near-Po sites have little bioturbation-generated irrigation, and solute transport back into water column is dominated by molecular diffusion. It appears that where sediment accumulation is very high, deep-burrowing benthos cannot survive well, whereas they can thrive sufficiently to generate abundant irrigation canals through the sediment where the accumulation rate is below a critical threshold. This is consistent with the pattern of endobenthos described above for the sites studied by Mancinelli et al. (1998) and Simonini et al. (2004).

Although pore-water chemistry was documented for only 2 cm depth for the site offshore of Ravenna shown in figure 7.6, Hammond et al. (1984) had earlier examined a core of much greater length. In that study they examined the flux of Radon 222 and found that it escaped into the water column at about four times the rate predicted by a diffusion model. They found that approximately the upper 25 cm was being actively burrowed by polychaetes, producing a mixed zone down to 20–30 cm depth, below which Radon 222 concentrations were consistent with predictions of the diffusion model. Polychaete irrigation of the upper 20–30 cm was vigorously contrib-

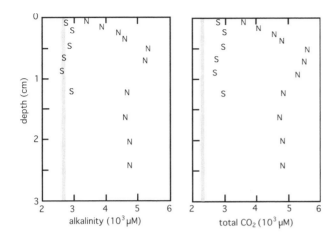

FIGURE 7.6 Two of the seven pore-water attributes determined by Hammond et al. (1999) for a 15 m–deep station close to the Po Delta (N; position marked on fig. 7.4 as H) and a 15 m–deep station (S) midway between Ravenna and Ancona. The vertical gray band at the left side of each diagram indicates alkalinity and CO_2 concentration in the overlying water.

Modified from Hammond et al. 1999.

uting to flux of nutrients from fairly deep within the sediments into the water column.

EASTERN BENTHIC ASSEMBLAGES

There is an ill-defined but apparently continuous zone with abundant epi-benthic suspension-feeders offshore of the Balkan coast of the Adriatic Sea, extending from the Gulf of Trieste in the north, southward off the coast of Istria, and to an unknown extent south of Istria (Fedra et al. 1976). Because of the position of active marine laboratories, it has been extensively studied with published results only for the Gulf of Trieste and for embayments and the open Adriatic near Rovinj, Croatia.

GULF OF TRIESTE

The Gulf of Trieste resembles a scaled-down, shallower version of the northern Adriatic Sea, rotated 90° clockwise. It has riverine influence on the north, opposite a carbonate coast along the Istrian coast on the south, with intermediate conditions along the east coast. Circulation below 10 m depth within the Gulf of Trieste is cyclonic (Stravisi 1983), an extension of the basic Adriatic-wide circulation pattern. Therefore, as the water passes Trieste and Monfalcone it encounters the first of the Italian nutrient-rich rivers before continuing further west and south along the Italian coast.

Much of the gulf was studied by Orel and Mennea (1969), who took 0.2 m² grab samples over a gridwork of more than one hundred stations, providing more detailed knowledge of the benthos than did Vatova's (1949) seven stations within the gulf. Aside from a biologically unstable, polluted zone adjacent to the city of Trieste, they recognized three groups of samples. The first, a narrow zone along the north shore, near the Isonzo Delta, was characterized almost exclusively by deposit- and suspension-feeding endobenthic polychaetes and bivalves, plus an epibenthic mud-grazing gastropod.

MID-GULF *SCHIZASTER* COMMUNITY

Most of the gulf was occupied by a benthic community with about 70 percent of the biomass comprised of endobenthos, most of which was deposit feeding (fig. 7.7). Over half of the biomass of deposit-feeders consisted of the burrowing echinoid *Schizaster canaliferus*. The sediment-penetrating carnivorous gastropod *Aporrhais pespelecani* and both endobenthic and epibenthic holothurians were also abundant in this central zone. The epibenthos (30 percent of total biomass) was almost entirely mobile, a mix of suspension-feeders, grazers, detritivores, and carnivores; only 2 percent of the biomass comprised epibenthic sedentary suspension-feeders. Sediment below 20 m depth in at least part of the area is anoxic except at the sediment–water interface, due to the oxygen demand of decomposing organic matter within it (Faganeli et al. 1991). Despite the within-sediment anoxic conditions, presence of abundant nutrients has resulted in the sediment being completely bioturbated down to depths of at least 25 cm (Faganeli et al. 1991).

Schizaster canaliferus is a spatangoid echinoid, and where spatangoids are common they can have a pronounced and complex influence on nutrient flux across the sediment–water interface (Lohrer et al. 2004). This is in part because of the effect that their burrowing and life processes have on microbes on and within the sediment. Their burrowing increases sediment porosity and therefore diffusion and advection of dissolved oxygen, and their defecation increases the supply of dissolved organic matter. These results enhance bacterial populations (part of the echinoids' food source), which regenerate dissolved inorganic nitrogen and phosphorus. During periods of illumination, those nutrients plus ammonia released by the echinoids are partially utilized by microphytes such as diatoms, which too are grazed by the echinoids but also release oxygen that is utilized by both echinoids and bacteria. The bioturbation- and photosynthesis-enhanced exchange of oxygen in the sediment contributes to binding phosphate onto clays so that it becomes refractory and no longer recycles through the ecosystem. *S. canaliferus* burrows just above the redox boundary, which in the silts of study sites in the Gulf of Trieste varied between 37 and 45 mm (Schinner 1993).

Pronounced seasonal change in bioturbation intensity in the mid-gulf is indicated by pore-water profiles (Cermelj et al. 1997). Winter sediment pore-water profiles show an exponential increase with depth within the sediment for ammonia, phosphate, silica, and dissolved inorganic carbon, whereas summer profiles have constant or only gradually increasing values with depth. This suggests that vigorous bioturbation, which one would

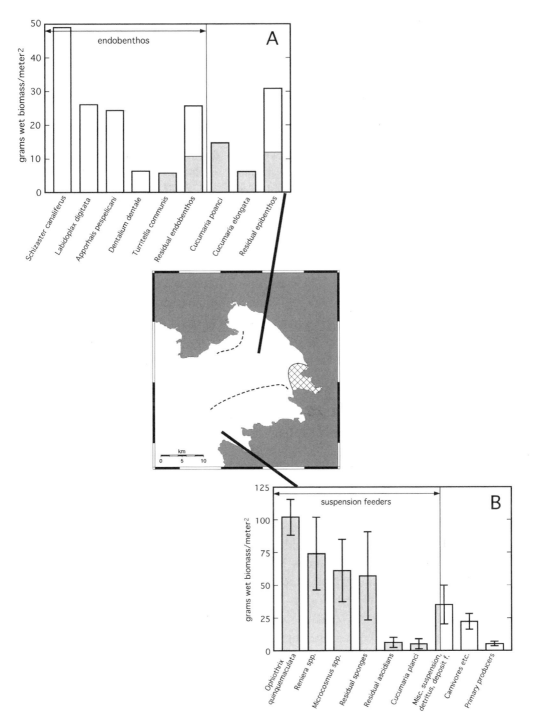

FIGURE 7.7 Benthos of the Gulf of Trieste. A small heavily polluted region (cross-hatched) adjoins Trieste, endobenthic bivalves and polychaetes characterize a narrow region along a delta on the northern coast, and otherwise a mixed fauna (A) is found in the mid-gulf, and a zone dominated by epibenthos (B) occurs along the Istrian coast. Shaded bars denote suspension-feeders. Part B shows the pre-1976 biomass distribution of the *Ophiothrix-Reniera-Microcosmus* community in the Gulf of Trieste.

Data in part A are from Orel and Mennea 1969 and data in part B are from Fedra et al. 1976. Base map generated by Online Map Creation (www.geomar.de/omc/omc_intro.html).

expect given the prevalence of endobenthos in the region (fig. 7.7A), happens during summer but not during winter. Indeed, laboratory experiments indicate that *Schizaster canaliferus* burrows twice as fast at 16–22°C as at 8–14°C, which encompasses the normal bottom-water temperature range in the Gulf of Trieste (Schinner 1993). In winter the pore-water chemistry fits a model for nutrient flux by diffusion rather than by a mix of diffusion and irrigation via biologically generated channels within the sediment. The variation from season to season mimics the north–south contrast seen in figure 7.6, but as a temporal rather than geographic pattern.

Different pore-water profiles in the mid-gulf between winter and summer are matched by opposite changes in the overlying water below the pycnocline for suspended matter, particulate and dissolved organic carbon, particulate and dissolved nitrogen, and orthophosphate. Concentrations of these materials increase to maxima above the bottom at the summer–autumn transition, which suggests vigorous microbial activity and resuspension from the sediment into water during summer (Faganeli et al. 1991).

Ophiothrix-Reniera-Microcosmus Community North of Istria

The third community in the Gulf of Trieste distinguished by Orel and Mennea (1969) occurred in a several km–wide band along the southern coast of the gulf and was largely epibenthic, dominated by suspension-feeders: an ophiuroid plus sponges and ascidians. They considered the third community to be an extension of the epibenthos-rich zoocoenoses described by Vatova (1935, 1949) along the open-Adriatic coast of Istria.

Offshore of Piran, Slovenia, at the northwestern tip of Istria, the benthic community characterized as epibenthos-dominated by Orel and Mennea (1969) has been continuously studied for several decades. The entire offshore area below 20 m depth was occupied by a single community characterized by the ophiuroid *Ophiothrix quinquemaculata*, a demosponge (*Reniera* spp.), and an ascidian (*Microcosmus* spp.) (fig. 7.7B), termed the ORM community based on the three characteristic taxa. There is an abrupt transition to a different, lower biomass community in shallower water (Fedra et al. 1976; Fedra 1977).

Slightly higher nutrients characterized the region occupied by the ORM community than occur in the open Adriatic off the Istrian coast, with steady currents of about 0.1 m sec⁻¹, visibility of about 1.5 m, and a sediment substratum of muddy sand. Samples of 0.25 m² were hand-collected by scuba divers as the primary basis for the study of the community, which probably causes the epibenthos to be underrepresented (Fedra et al. 1976). Although the community was clearly epibenthos-dominated, the percentages given below likely would have a somewhat higher proportion of endobenthos had Van Veen grab samples been the basis for the study.

Macrobenthos of the ORM community averaged about 370 g m^{-2} wet biomass during the years of study (Fedra et al. 1976). The three characteristic taxa of the community are suspension-feeders, and on average they constituted 67 percent of the wet biomass, with all suspension-feeders in the community comprising 87.5–89 percent of the wet biomass (fig. 7.7B). The community components exhibited similar proportions to one another across the entire area and were concentrated in clumps growing on local hard substrata such as mollusk shells lying on—or being transported by hermit crabs across—the sediment surface (Fedra et al. 1976).

Hermit crabs living in gastropod shells in the Gulf of Trieste and elsewhere maintain the shells above the sediment–water interface longer than they ordinarily would be after the death of the original gastropod inhabitant. In addition, they potentially can carry the shells onto sediment surfaces where shell litter may not otherwise be present. Thus hermit crabs may bring a hard substratum on which organisms already are growing and eventually deposit the shell and its accumulated epibionts, both the attached sedentary organisms and the mobile organisms living on and within the attached organisms. Such epibiotic proliferation may ultimately be sufficiently large that when the shell is abandoned or its inhabitant expires, it is too heavy to attract another hermit crab but continues to grow and recruit new settlers in situ, forming a local epibenthic patch on an otherwise uninhabited sediment surface. This is an important process for maintenance of the epibenthos in the Gulf of Trieste, where over 120 species of epibionts and endolithic organisms compete for a place on the shells (Stachowitsch 1980).

Analysis of energy flow within the ORM community is rather interesting from a paleontological perspective. Biomass, number of individuals, caloric values, and respiration rates per m^2 were determined for the most abundant 50 megascopic species in the community (Fedra 1977). Good correspondence was found in lognormal cumulative curve shape between wet biomass, caloric values, and respiration rates, although the curve based on number of individuals approaches a logarithmic curve. Compared to respiration rates, the curves for biomass and caloric curves overestimate the importance of large, less common species, whereas number of individuals overestimates the importance of species with numerous, small individuals. That is more or less what one would expect, but an interesting result for paleontologists is that when the species are grouped as in figure 7.7B, "the numbers of individuals were found to provide a better estimate of the respective contributions to the community's energy flow than the applied biomass measures" (Fedra 1977:239).

Ophiothrix quinquemaculata dominated the community in both numbers and biomass, ranging from 50 to 250 individuals m^{-2} (Fedra et al. 1976). Individuals of *O. quinquemaculata* perched almost exclusively on several species of sponges (Fedra 1977). *O. quinquemaculata* and the less abundant suspension-feeding holothurian *Cucumaria planci* are rheophiles, orienting their feeding structures to intercept the ambient current and even migrating upcurrent (Ölscher and Fedra 1977). Although some adult *O. quinquemaculata* were found resting directly on the sediment, all the juveniles and most adults were on elevated substrates, which increases their exposure to ambient currents above the reduced-velocity boundary layer that exists just above the sediment surface.

Nutrient transfer at the sediment–water interface is an expression of the sources of nutrients to the benthos. In the Gulf of Trieste, almost all the carbon and nitrogen (95 percent and 80 percent, respectively) are recycled at the sediment–water interface, although at least half the phosphate and biogenic silica are lost to the sediments (Faganeli et al. 1991; Bertuzzi et al. 1997). The released nitrogen, silica, and phosphorus appear to be utilized by benthic microalgae, including diatoms, while the surplus of released dissolved inorganic carbon is available to the phytoplankton below the pycnocline. This implies that during the time that the water column is stratified and biological processes are most vigorous, the benthic communities and near-bottom water are interrelated with one another but are relatively independent of the above-pycnocline water mass. The pelagic nutrients above the pycnocline are largely introduced in river runoff rather than being recycled benthic nutrients (Bertuzzi et al. 1997). Sedimentation from the pelagic realm normally adds a modest amount to the benthic nutrient supply, over the long-term essentially replacing the nutrients lost to the sediment.

Having existed at least for decades, the ORM community offshore of Piran was considered to be stable and long term; in fact, it had existed there ever since the area had first been studied (Fedra et al. 1976). In September 1983 it was devastated by a severe anoxic event (Stachowitsch 1984; Faganeli et al. 1985). Absence of strong winds and dry weather through the summer generated prolonged existence of the pycnocline. In September, decomposing organic matter that had been hung up on the pycnocline finally penetrated to the seafloor, consuming available oxygen and killing and causing the decay of the sedentary and poorly mobile macrobenthos. The oxygen demand generated by the decay was sufficiently strong to cause sulphate reduction, generating H_2S (Faganeli et al. 1985). In other words, the large short-term input of nutrients from the pelagic realm abruptly disturbed the normal relative independence of the pelagic and benthic realms.

Michael Stachowitsch photographically documented death of the benthos during the anoxic event of September 1983 and as the subsequent mass mortality of the ORM community proceeded over the next two weeks. He produced (1984) a series of fascinating photographs with an accompanying

TABLE 7.1

Benthic Organisms Affected by September 1983 Anoxic Events
in the Gulf of Trieste

Benthic fish	Gaping mouths; dead on first day (*Bobius jozo*).
Crustacea	Entangled in marine snow on clumps (*Pisidia longicornis*, shrimp, *Pilumnus spinifer*); emergence from burrows (*Upogebia tipica*, *Jaxea nocturna*, *Axius stirhynchus*); in humped position on sediment, walking on or swimming above bottom (*Squilla mantis*); aggregation on mounds, extension from shell, dead next to shell (*Paguristes oculatus*); abandonment of invertebrate host (*Pinnoteres pinnoteres*).
Holothuroidea	Epibenthic forms: evisceration (*Holothuria tubulosa*); endobenthic forms: emergence and evisceration (*Thyone fusus*, *Labidooplax digitata*).
Ophiuroidea	*Ophiothrix quinquemaculata*: interruption of suspension feeding, abandonment of multispecies clumps, flat posture on sediment surface, coiled arms, fragmentation, overturning; *Ophiura texturata*: humped position with elevated central disc; *Amphiura chiajei*: aggregation on mounds.
Asteroidea	*Astropecten* sp.: distended, gas-filled central disc; overturning.
Echinoidea	Emergence from sediment (*Schizaster canaliferus*).
Polychaeta	Emergence, moribund on sediment, fragmentation (*Eunice aproditois*).
Sipunculida	Emergence (early in *Golfingia elongata*, not until tenth day for *Sipunculus nudus*).
Gastropoda	Emergence from sediment (*Aporrhais pes-pelecani*).
Bivalvia	Emergence from sediment, gaping valves, extended siphons (*Cardium* sp., *Solecurtus* sp.).
Cnidaria (anemones)	Retraction of tentacles (*Cerianthus* sp.); detachment from hermit crab–occupied shells (*Calliactis parasitica*); constricted columns, elevation of tentacle crown, emergence (exposed pedal discs), discharged acontia.

Organisms are listed approximately in the sequence in which their stress behaviors and mortality developed over a three-day period.
From Stachowitsch 1984.

description of the destruction of the community. Effects were noticed immediately in benthic fish, followed by an unfolding sequence of stress behaviors and death of epibenthos and endobenthos (table 7.1). Organisms in all megaguilds were affected, but the settling mucilaginous masses that initiated the anoxic event clogged the feeding apparatus of suspension-feeders, for which the masses are particularly toxic (Müller et al. 1998).

Recovery from the 1983 anoxia-induced mass mortality has been irregular and faltering (Stachowitsch 1991, 1992; Kollmann and Stachowitsch 2001), demonstrating how precarious are such communities based on sedentary suspension feeding. During the decade following the 1983 event, wet biomass never exceeded 50 percent of the pre-1983 average, and the previously dominant ophiuroids, sponges, and ascidians fared even worse (fig. 7.8). At the same time there has been a prolonged increase in omnivorous, mobile hermit crabs (Kollmann and Stachowitsch 2001). Subsequent smaller-scale stresses on the benthos have had disproportionate effects: commercial trawling for scallops in 1987, a local and relatively minor anoxic event in 1988, and trawling again in 1990. Over a nineteen-day period in 1990, commercial trawling reduced coverage of the seafloor by epibenthos from 6 percent to 3.5 percent (Kollmann and Stachowitsch 2001).

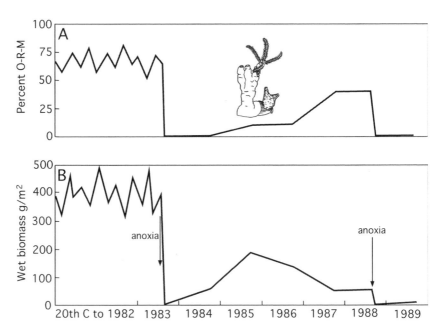

FIGURE 7.8   Lingering effects of the 1983 anoxic event in the Gulf of Trieste. A) Percent of wet biomass comprised of *Ophiothrix quinquemaculata* (O), *Reniera* spp. (R), and *Microcosmus* spp. (M). B) Wet biomass in grams per m².

Modified from Stachowitsch 1991, 1992.

A remarkably similar change occurred farther south off the Istrian coast, triggered by a widespread anoxic event in 1989 (Jaklin 2002). Recovery has been very slow, and five years after the event, the pre-anoxia communities dominated by sedentary suspension-feeders had not yet recovered. The region was largely populated by endobenthic deposit-feeders, omnivores, and carnivores, although the original suspension feeder–based community was beginning to return in at least one area by 1997 (see below on Sv. Ivan).

MID-ISTRIAN OFFSHORE

The Adriatic benthos off the Istrian coast in the vicinity of Rovinj has been studied more intensely than in most other areas of the Adriatic. Vatova's extensive survey of the benthos (1935) covered an area about 15 by 15 km, with Rovinj midway up the landward margin (fig. 7.9). Numerous subsequent studies of benthos within the area have used various techniques to address different sorts of questions, and several of the studies are particularly pertinent here. Intra-annual variation in biomass and community structure was tracked by Gamulin-Brida et al. (1968) and Zavodnik (1971); Seneš (1988a–c, 1989) used photo-transects and qualitative samples by divers to document changes in benthos across depth and substratum types; and McKinney and Jaklin (2000, 2001) and McKinney (2003) examined some of the bryozoan-rich localities.

A several kilometer–wide band of largely relict sand lies offshore of the Istrian coast (fig. 7.9), bounded to the west by a broad zone of progressively muddier sediments all the way to the foot of the Po Delta (fig. 6.7) and to the east by a narrow silt-dominated belt between the sand tract and the shore. Sediment along the highly indented coast ranges from clay-rich muds within the most sheltered areas such as Lim Channel (Canal di Leme) to limestone cobbles at the foot of coastal promontories.

Samples collected by Vatova (1935) give a basic overview of the benthos offshore of Rovinj. They showed no difference in number of species per sample for any of the sedimentary environments (table 7.2), grouping the complex of shore-zone samples together. However, biomass was highest within the near-shore silty mud belt. The sedentary epibenthic portion of the benthos also had its highest biomass per sample in the near-shore silty mud belt, where it made up about two-thirds of the total biomass collected in the 28 samples within this zone. (This is reflected in the mean values rather than in the median values for total biomass and for epibenthic biomass.) Although biomass of the sedentary benthos declined in the offshore sand belt, it still comprised over 50 percent of the total.

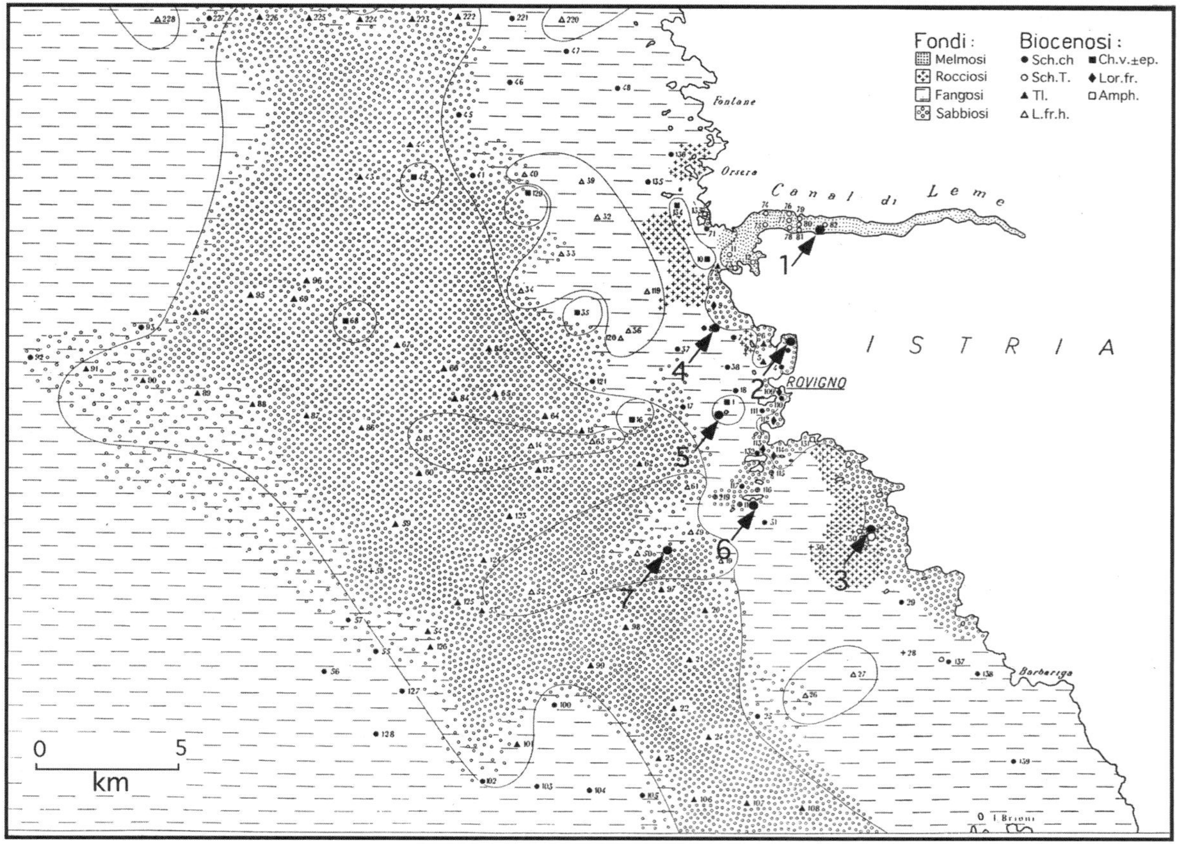

FIGURE 7.9 Vatova's 1935 map of benthic assemblages (zoocoenoses) and substrata offshore of Rovinj, Croatia. Patterns represent sediment type: melmosi (fine dots in Canal di Leme) = organic rich mud; rocciosi (crosses in local patches along shore) = carbonate pebbles to boulders; fangosi (dashes near shore and in westernmost area) = sandy silt-dominated mud; sabbiosi (small circles in northwest–southeast median strip and locally along shore) = carbonate sand. Fine solid lines approximate the boundaries between Vatova's assemblages. Named and numbered arrows indicate additional sites that have been subsequently studied in more detail (see below). Arrowed sites are numbered as in the text section on mid-Istrian offshore site studies.

TABLE 7.2 General Biomass Characteristics of Vatova's (1935) Samples of the Benthos Organized in Shore-Parallel Sedimentary Belts Offshore of Rovinj, Croatia

SEDIMENTARY BELT	METRIC	NUMBER OF SPECIES	TOTAL WET BIOMASS (g 0.2 m²)	BIOMASS OF ENDOBENTHOS	BIOMASS OF SEDENTARY EPIBENTHOS
Shore-zone	Range	3–18	0.4–159.3	0.3–121.2	0–133.0
clay-rich mud,	Median	10.5	16.2	6.0	0
sand, rocks (22)	Mean	11.0±4.9	34.4±42.2	17.7±27.0	9.5±29.7
				(51.3%)	(27.5%)
Near-shore	Range	3–27	3.6–500.8	0.1–84.0	0–500
sandy silt (28)	Median	7	29.9	6.4	13.2
	Mean	8.8±5.3	68.8±99.8	13.2±18.3	43.8±96.7
				(19.2%)	(63.7%)
Offshore	Range	2–25	0.3–352.1	0.2–112.3	0–289.1
sand (60)	Median	8	9.7	2.8	1.2
	Mean	9.4±6.0	47.2±83.9	9.5±17.9	28.9±64.5
				(20.2%)	(61.3%)
Offshore	Range	3–17	0.9–47.8	0.2–25.8	0–25.0
sandy silt (17)	Median	8	14.5	2.7	0
	Mean	8.3±4.2	13.5±12.7	6.9±8.6	3.6±7.2
				(51.3%)	(27.1%)

Values given are mean ± one standard deviation; number of samples for each sedimentary belt is given in parentheses in the left column.

Note that the median and means for species number are very close, suggesting normal distribution of species numbers within each belt and actually for the region as a whole. This contrasts conspicuously with all the biomass values, in which only one set of biomass values (total biomass in the offshore sandy silt) have nearly identical median and mean values with standard deviation less than the calculated mean. Highly developed patchiness of distribution is reflected in all the other biomass cells in table 7.2, with the mean value being conspicuously greater than the median and with very high standard deviations; in fact, the median sedentary epibenthic biomass is 0 in both the shore zone and the offshore sandy silt belt. The most basic patterns seen in table 7.2 are that sedentary epibenthos makes up over half the total benthic biomass in the two central belts even though it is very patchily distributed, and that although endobenthos is present in all 127 samples taken, it makes up the majority of the biomass only in the inner and outer belts. These sediment-belt related patterns can be supplemented by different groupings of the samples.

Vatova grouped the samples of benthos offshore of Rovinj into seven zoocoenoses ("biocenosi" in fig. 7.9; table 7.3). He determined that one zoocoenosis (*Schizaster chiajei*) occupied most of the area of the two silty mud belts and a second (*Tellina*) occupied the majority of the offshore sand belt. Both of these zoocoenoses have consistently low species richness. The highest average species richness characterizes the *Chione verrucosa* zoocoenosis. Three others are intermediate in species richness: the *Loripinus fragilis* and *Amphioxus* zoocoenoses, which are part of the along-shore benthos, and the *Lima hians* zoocoenosis. The *Lima hians* and *Chione verrucosa* zoocoenoses were encountered locally across the inner silty mud and the offshore sand belts. These are the two zoocoenoses with the highest average total biomass and the highest sedentary epibenthic biomass, which on average constitutes about 70–75 percent of the total in the two zoocoenoses. Although they are both patchily distributed, they most commonly occur near the transition (roughly 4 km offshore) between the inner silty mud and outer sand belts.

If Vatova's samples are organized into 2 km–wide shore-parallel bands (further dividing the onshore 2 km into two 1 km–wide bands), a more striking pattern emerges (table 7.4). Number of species, total biomass, and sedentary epibenthic biomass are maximum at 2–6 km offshore, which tracks the boundary between the near-shore silty mud and offshore sand belts. Endobenthic biomass shows an almost monotonic decrease from 1–2 km offshore to the farthest offshore zone.

Biomass of sedentary epibenthos increases steadily offshore to the 2–6 km belts (highest median at 2–4; highest mean at 4–6), after which it irregularly falls off (table 7.4). Interaction of this distribution with the steady decline of endobenthos offshore results in epibenthos comprising a very large proportion of benthic biomass from 1 to 12 km offshore, with particularly high proportions of sedentary epibenthos 4–6 km and 8–10 km offshore (fig. 7.10). A precipitous drop in epibenthos into the outermost two zones makes the small volume of endobenthos there constitute from one-third to two-thirds of the total biomass, with virtually all the other biomass occurring in *mobile* epibenthos. The outermost zone is the first zone from the land in which endobenthos averaged more than 50 percent of the benthic biomass.

The epibenthos in Vatova's samples from 2 to 12 km offshore consistently was dominated by ascidians, largely *Microcosmus* spp.; from 2 to 6 km sponges and epibenthic holothurians codominated with *Microcosmus*, supplemented by epibenthic attached and free-lying bivalves (fig. 7.11). The endobenthos was dominated by the echinoid, *Schizaster canaliferus*, with moderately abundant endobenthic bivalves near-shore and moderately abundant polychaetes farthest offshore (fig. 7.11). Carnivorous within-sediment-feeding gastropods such as *Aporrhais pespelecani* reach their

TABLE 7.3 General Biomass Characteristics of Vatova's (1935) Zoocoenoses Offshore of Rovinj, Croatia

Vatova Zoocoenosis	Metric	Number of Species	Total Wet Biomass (g 0.2 m²)	Biomass of Endobenthos	Biomass of Sedentary Epibenthos
Schizaste–chiajei	Range	3–20	2.1–500.8	0.1–121.2	0–500.0
(40)	Median	8.5	27.7	16.9	0.3
	Mean	9.0±4.6	51.3±81.4	20.8±25.6	23.1±79.5
				(40.5%)	(45.0%)
Schizaster–	Range	3–7	4.2–7.2	4.2–7.2	0
Turritella (2)	Median	5	5.7	5.7	0
	Mean	5	5.7	5.7	0
				(100%)	(0%)
Tellina (44)	Range	2–12	0.3–145.3	0.2–9.4	0–21.0
	Median	6	3.4	1.3	0
	Mean	6.4±2.9	9.3±22.0	2.1±2.1	1.8±3.8
				(22.1%)	(18.8%)
Lima hians (20)	Range	5–25	6.1–352.1	0.6–65.1	0–234.0
	Median	11	30.8	6.2	22.2
	Mean	12.6±5.9	71.0±84.9	9.8±14.2	50.3±60.0
				(13.7%)	(70.1%)
Chione	Range	4–27	1.2–346.7	0.4–84.0	0–289.1
verrucosa (11)	Median	15	149.3	21.9	130.6
	Mean	15.2±7.8	147.1±120.1	22.2±23.9	108.1±103.0
				(15.1%)	(73.5%)
Loripinus	Range	9–18	4.1–58.8	1.3–47.2	0–38.5
fragilis (7)	Median	13	48.6	2.7	0
	Mean	13.4±3.9	32.5±25.6	9.7±16.7	5.6±14.5
				(29.8%)	(17.2%)
Amphioxus (3)	Range	7–18	0.8–4.9	0.8–4.9	0
	Median	10	2.4	2.2	0
	Mean	11.7±5.7	2.7±2.1	2.6±2.1	0
				(97.5%)	(0%)

Values given are mean ± one standard deviation; number of samples per zoocoenosis are given in parentheses in the left column.

TABLE 7.4 General Biomass Characteristics of Vatova's (1935) Samples of the Benthos Organized in Incremental Shore–Parallel Zones Offshore of Rovinj, Croatia

DISTANCE FROM SHORE/ DEPTH	METRIC	NUMBER OF SPECIES	TOTAL WET BIOMASS (g 0.2 m²)	BIOMASS OF ENDOBENTHOS	BIOMASS OF SEDENTARY EPIBENTHOS
0–1 km /	Range	3–18	0.4–159.3	0.1–121.2	0–133.0
17.2±9.3 m	Median	9.5	18.4	5.3	0
(24)	Mean	9.6±4.9	42.4±48.9	17.9±27.2	12.0±31.0
				(42.2%)	(28.3%)
1 to 2 km /	Range	3–27	2.4–149.3	1.8–84.0	0–130.6
26.4±9.1 m	Median	8.5	26.8	10.9	0.1
(10)	Mean	10.7±7.8	42.7±54.0	18.7±24.8	18.5±40.4
				(43.8%)	(43.3%)
2 to 4 km /	Range	2–25	0.6–352.1	0.6–65.1	0–234.0
33.0±2.7 m	Median	11	30.3	7.9	16.1
(15)	Mean	11.2±5.8	64.2±91.6	13.3±17.1	29.7±61.2
				(20.8%)	(46.2%)
4 to 6 km /	Range	3–25	1.2–500.8	0.3–36.2	0–500.0
33.1±1.9 m	Median	10	37.1	5.0	10.2
(19)	Mean	11.7±6.9	88.6±133.0	10.1±11.8	66.9±120.5
				(11.4%)	(75.5%)
6 to 8 km /	Range	3–12	0.9–145.3	0.4–28.8	0–123.1
34.5±3.0 m	Median	9.5	20.4	3.5	0.1
(16)	Mean	8.4±3.1	33.5±43.8	6.6±7.9	13.0±33.4
				(19.9%)	(38.8%)
8 to 10 km /	Range	2–23	0.9–312.3	0.2–22.6	0–289.12
35.4±3.9 m	Median	6.5	5.6	1.8	3.8
(18)	Mean	7.8±5.0	29.8±72.8	4.3±6.4	24.7±67.4
				(14.4%)	(83.0%)
10 to 12 km /	Range	2–17	0.3–243.8	0.3–112.3	0–225.0
38.4±3.6 m	Median	7	7.3	4.2	0
(11)	Mean	8.7±5.7	42.7±74.6	18.5±32.4	22.2±57.4
				(43.2%)	(51.9%)
12 to 14 km /	Range	3–12	0.7–22.0	0.4–8.0	0–0.1
37.4±2.0 m	Median	7	4.8	1.7	0
(8)	Mean	7.6±3.4	8.4±8.2	2.6±2.7	0.01±0.02
				(31.0%)	(0.1%)
Over 14 km /	Range	3–10	0.9–24.7	0.2–22.1	0
35.7±0.8 m	Median	6	2.0	1.0	0
(6)	Mean	6.3±2.8	7.6±9.9	5.4±8.6	0
				(71.4%)	(0%)

Values given are mean ± one standard deviation; number of samples per zoocoenosis are given in parentheses in the left column.

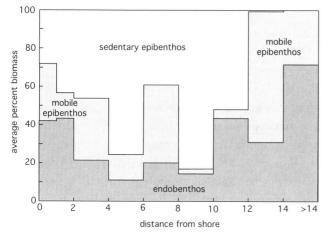

FIGURE 7.10 Percent wet biomass of endobenthos, mobile epibenthos, and sedentary epibenthos collected in grab samples indicated in fig. 7.9, organized in shore-parallel bins. Note the prevalence of epibenthos in all but the zone greater than 14 km offshore.

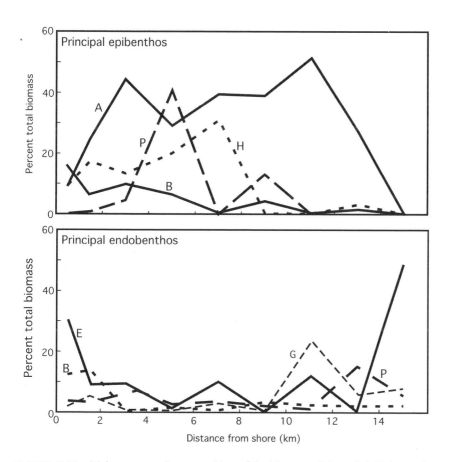

FIGURE 7.11 Major taxonomic composition of the biomass of Vatova's (1935) samples of the benthos from onshore to offshore of Rovinj, Croatia. Principal epibenthos include Ascidia (A), epibenthic Bivalvia (B), epibenthic suspension-feeding Holothuroidea (H), and Porifera (P). Principal endobenthos include endobenthic Bivalvia (B), Echinoidea (E), and Polychaeta (P); and endobenthic and sediment-probing Gastropoda (G).

highest abundance in the farthest offshore belts, although they are also moderately common within 2 km of the shore.

Dredging at twenty stations offshore of Rovinj during the mid-1960s yielded a higher species richness per locality than Vatova had encountered (Gamulin-Brida et al. 1968). This difference may be due entirely to the different sampling technique used by Gamulin-Brida et al., who pulled a dredge that bites a few cm deep into the sediment across the occupied stations and thereby sampled a much larger area than the 0.2 m^2 area encompassed by Vatova's Van Veen grab. They determined that there is a complex of local communities within the inner silty mud belt, most dominated by sedentary suspension-feeders, with the suspension-feeding ophiuroid *Ophiothrix quinquemaculata* locally abundant. The highly variable diversity of local communities within this belt is consistent with Vatova's more general observations, with the complex of local environments that exist near-shore, and with the patchiness in distribution of biomass indicated by the statistical characteristics given in table 7.2.

Gamulin-Brida et al. (1968) confirmed that the transition from the inner silty mud belt to offshore sand belt had higher species richness than closer to shore and that it was highly dominated by suspension-feeding epibenthos, including areas with remarkably abundant *Ophiothrix quinquemaculata*. *O. quinquemaculata* occurs locally in densities up to 100 m^{-2}, and as in the Gulf of Trieste they tend to climb up on promontories such as sponges and ascidians, although photographs show them living directly on the sediment in large numbers, with arms bent up into the water column (fig. 7.12; Czihak 1959).

Along the boundary between the near-shore silty mud belt and the offshore sand belt Gamulin-Brida et al. (1968) reported extensive development of abundant erect calcified bryozoans and in other areas abundant rhodophyte algae. The erect bryozoans exhibited the same range of diverse growth habits that were abundant in the Paleozoic: fenestrate, branched bifoliate, sheetlike bifoliate, articulated small cylindrical branched, and even robust nonarticulated branched colonies (fig. 7.13). Curiously, the dense growths of erect calcified bryozoans did not show up in Vatova's 1935 tables of data for the Rovinj region. He did, however, have several stations in the region where the erect bryozoans are abundant, marked by an undefined symbol (an encircled cross, such as station 219 west of arrowed locality 6 in fig. 7.9) that did not appear in any of the tables of data. Having worked with this material, I can imagine that he encountered such an inextricably tangled morass of so many erect and tightly encrusting species in the grab samples that he decided in the end that it was impossible to determine biomass or even to count the individuals per species.

It is clear that the region offshore of Rovinj has a more complex pattern of distribution of benthic organisms than within the Gulf of Trieste (fig. 7.7).

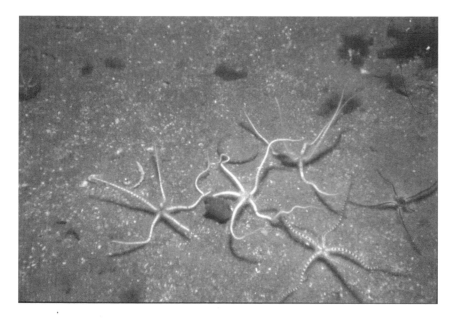

FIGURE 7.12   Muddy sand seafloor 5 km off Pula, Croatia, 32 m depth, with several *Ophiothrix quinquemaculata* feeding.

Photograph courtesy of A. Jaklin.

Just as offshore of Piran, the benthos at >20 m depth offshore of Rovinj is dominated by sedentary and mobile epibenthic suspension-feeders (table 7.4). However, in contrast with the uniformity of dominance of three suspension-feeding taxa (*Ophiothrix, Reniera,* and *Microcosmus*) across the area studied by Fedra et al. (1976), ascidians—largely *Microcosmus*—were the only consistently abundant taxon across the area. Vatova found sponges and epibenthic suspension-feeding holothurians to be abundant locally (fig. 7.11) but found *Ophiothrix quinquemaculata* to constitute over 25 percent of the biomass in only two samples (Vatova 1935). The absence of abundant *O. quinquemaculata* in most of Vatova's transition and sand belt samples may be the result of a pressure pulse from the descending Van Veen grab pushing away the almost neutrally buoyant individuals. *O. quinquemaculata* feed almost continuously (Ölscher and Fedra 1977) and when feeding on bare sediment are vulnerable to physical disturbance because most of their arm length is extended up into the water to intercept plankton (fig. 7.12). Otherwise they generally move slowly and respond to physical disturbance by clasping any projecting substratum more tightly.

Alternatively, their relative paucity in Vatova's samples may be due to pronounced patchiness of distribution on various scales from some areas of high abundance on local clumps as offshore of Piran, and elsewhere aggregating into local more uniform high abundance across areas of a few tens of meters. In four localities in the offshore sand and in the inner silty mud to

FIGURE 7.13 Common erect and free-living calcified bryozoans offshore of Rovinj, Croatia.

A) *Cellaria fistulosa*; articulated, narrow cylindrical branches, colony approximately 15 cm diameter. B) *Myriapora truncata*; continuously calcified robust cylindrical branches, colony approximately 8 cm across, partially encrusted by sponge. C) *Adeonella pallasii*; bifoliate branches, colony approximately 7 cm across. D) *Pentapora fascialis*; broad bifoliate sheets and branches, approximately 10 cm width shown. E) *Reteporella grimaldii*; unilaminate branches in fenestrate colony, about 3 cm width shown. F) *Rhynchozoon revelatus*; free-living domal colony, about 4 cm in diameter.

All photographs are of live colonies, A–E in situ, with feeding structures (lophophores) extended and causing colony locally to appear fuzzy or out of focus. Light-colored spots on *R. revelatus* are crests of small promontories (monticules) where lophophores are absent and filtered water flows away from the colony surface.

A–E courtesy of A. Jaklin.

sand belt transition, Gamulin-Brida et al. (1968) had from 800 to 5,000 *O. quinquemaculata* per dredged locality, constituting from over 60 percent to about 90 percent of the total number of individual organisms.

LOCAL DISTRIBUTION PATTERNS

Jan Seneš, a prominent Slovakian geologist, became fascinated with the distribution of eastern Adriatic benthos and its residual skeletal accumulations. He was particularly interested in its potential usefulness for interpretation of biofacies in Cenozoic deposits of southern and central Europe. Between 1964 and 1974 he studied the living communities and sediments along approximately 15 km of scuba transects in the vicinity of Rovinj. Along the length of the transects, Seneš and his team took numerous sediment samples, determined the visible taxa within m² quadrats, and accumulated notes as they swam above the seafloor along the transect tract. These studies resulted in a series of papers in which the composition and distribution of benthic communities and the associated substrate types were described qualitatively (Seneš 1988a–c, 1989).

Several carbonate islands lie offshore and south of Rovinj (fig. 7.9), and at least one end of each of Seneš' transects extends to the supratidal zone of one of them. The general pattern is that the intertidal and shallow subtidal surfaces are steep rock faces, with a descending series of biological zones in which the proportion of encrusting sponges, cnidarians, bryozoans, and other animals increases, with a correspondingly decreasing proportion of soft and coralline algae. The lower slopes of the islands have progressively thicker sediment drapes over ledges until the interisland sediment abruptly appears at 20–25 m depth for some of the islands closest to the mainland but more typically at 30–35 m depth.

Seneš' transects across the sediment substrate in the vicinity of Rovinj (Seneš 1988a, c) were largely within an area characterized by Vatova (1935) as the *Schizaster–chiajei* zoocoenosis. Seneš recognized seven recurring biological "facies" on the marine sedimentary substrata (table 7.5; fig. 7.14), including local development of *eucoraligene* in deeper regions near Rovinj and farther south (fig. 7.14: no. 8). The local areas designated as eucoraligene comprise dense growth of encrusting and basally attached erect calcified coralline algae and bryozoans, forming a continuous lithified body. Individual eucoraligene areas may conceivably cover and obscure small underlying exposures of bedrock (and therefore not belong within benthic communities developed on a soft sediment substrate) or may develop by accretion from originally individual skeletons lying on unconsolidated sediment.

Four out of the seven biological groups determined by Seneš on sediment substrata are dominated by epibenthic suspension-feeders. Another is inhabited by forests of *Spirographis spallanzanii*, a suspension-feeding

TABLE 7.5   Dominant Organisms Indicated by Seneš (1988b, 1989) for
Soft-Sediment Biological Facies Crossed by SCUBA Transects Offshore of
Rovinj, Croatia

FACIES	DOMINANT	
C1a / ~30–35 m	Br	*Pentapora foliacea* (epibenthic suspension-feeder)
	Op	*Ophiura albida* (epibenthic mixed feeder)
C1b / ~45 m	Br	*Scrupocellaria reptans* (epibenthic suspension-feeder)
	Br	*Porella cervicornis* (epibenthic suspension-feeder)
	Br	*Hornera frondiculata* (epibenthic suspension-feeder)
C1c / ~30–35 m	Bi	*Arca noae* (epibenthic suspension-feeder)
C1d / ~30 m	Po	*Spirographis spallanzanii* (endobenthic suspension-feeder)
C1e / ~30–35 m	Bi	*Chlamys opercularis* (epibenthic suspension-feeder)
	Op	*Ophiothrix quinquemaculata* (epibenthic suspension-feeder)
C4 (eucoraligene)	Rh	*Pseudolithothamnion expansum* (photosynthesizer)
	Br	Diverse bryozoans (epibenthic suspension-feeders)
IC1 / ~25–30 m	Ga	*Aporrhais pespelicani* (endo-/epibenthic carnivore)
	Sc	*Dentalium rubescens* (endobenthic carnivore)

Bi = Bivalvia; Br = Bryozoa; Ga = Gastropoda; Op = Ophiuroidea; Po = Polychaeta; Rh = Rhodophyta; Sc = Scaphopoda

polychaete that inhabits tubes that are partially within the sediment but that also project several centimeters above the sediment–water interface, with the radially symmetrical feeding apparatus flared from the top of the tube. Only the shallowest, shoreward edge of the sedimentary interisland plain does not have epibenthic or partially emergent suspension-feeders as the most or next-most dominant member.

There is only a partially overlapping list of taxa that Vatova (1935), Gamulin-Brida et al. (1968), and Seneš (1988a–c, 1989) considered characteristic or even mentioned for the benthic biological groupings that each recognized in the area offshore of Rovinj. In part the differences can be

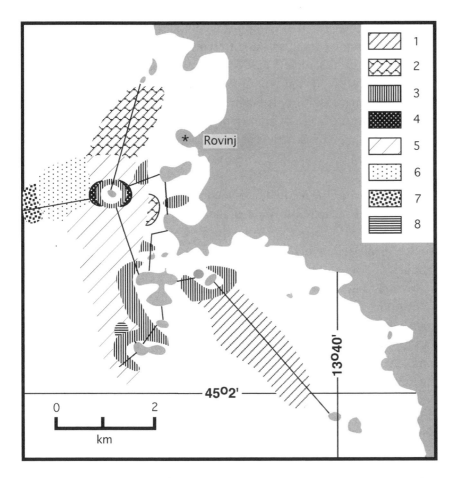

FIGURE 7.14 Paths of the major transects and distribution of Seneš' (1988a, 1989) biological facies on the sedimentary plain offshore of Rovinj, Croatia. Modified from Seneš 1988a. Unpatterned area along the coast is occupied by shallower, coastal biological associations; unpatterned areas offshore lie too far away from the transects and were not included in Seneš' study.

The facies are named after: 1: the gastropod *Turritella* and the bivalve *Aloidis*; 2: the gastropod *Aporrhais pespelicani*; 3: the bivalve *Pecten jacobaeus*; 4: the anthozoan *Cladocora cespitosa*; 5: the bryozoan *Pentapora fascialis*; 6: the bivalve *Arca noae* and the bryozoan *Pentapora foliacea*; 7: the bivalve *Chlamys opercularis* and the ophiuroid *Ophiothrix quinquemaculata*; 8: eucoraligene development.

attributed to three different sampling techniques, which would encounter patchily distributed organisms in different ways for each of the three techniques used (grab samples, dredging, and straight-line visual observation). The most important point to be made here is that all three studies produced similar results in that epibenthic suspension-feeders were recognized to constitute the most important benthos across the entire mud- and sand-covered plain from depths of about 25 m just offshore, across a broad area to the west before endobenthos became prevalent beyond the offshore sand belt.

Perhaps an argument could be made that the different taxa emphasized by Vatova (1935), Gamulin-Brida et al. (1968), and Seneš (1988a–c, 1989) reflect a several-decade change in the benthos. But such an argument would be more convincing with a temporal series of similarly collected samples. For one thing, as noted by Zavodnik (1971), the benthos offshore of Rovinj is much more species-rich than one would infer from the data of Vatova. This is almost to be expected in that, excluding the densely spaced grabs within the narrow estuary of Lim Channel, Vatova (1935) recorded data from about 130 grab samples covering 0.2 m² each, i.e., a total area sampled of little over 25 m². This was a Herculean effort in documenting the number and biomass of all epibenthic and endobenthic species within the 130 samples, but insufficient as a means of cataloguing the biodiversity of the region. For example, none of the surveys lists more than 40 sponge, 17 bryozoan, and 35 decapod crustacean species for the area, whereas subsequent taxon-specific studies have listed offshore of Rovinj 139 sponge species (Müller et al. 1984), 103 bryozoan species (Hayward and McKinney 2002), and, just within the diminishing eelgrass beds, 52 decapods (Števčić 1991).

SITE STUDIES

The benthic ecosystem in the vicinity of Rovinj is remarkably richly textured, even more so than indicated by Seneš' relatively fine-scaled organization into biological facies. This section summarizes seven studies at specific sites that give another perspective of the onshore–offshore increase in sedentary epibenthic suspension-feeders (fig. 7.9; table 7.6). Four of the sites were studied by Zavodnik (1971), who sampled each at monthly intervals for a year, using a Charcot-type dredge, which samples epibenthos well but cuts only about 2 cm deep into the sediment. He found pronounced seasonality in the number of individuals and in biomass for each of the major taxa at each site, with greatest lushness generally in late spring and early summer. That intra-annual variation is not portrayed here and instead the

TABLE 7.6   Proportions of Endobenthos and Attached Epibenthos from Adriatic Localities Offshore of Rovinj, Croatia

| | ONSHORE | | | OFFSHORE | | | |
Community	*Schizaster–Turritella*	*Amphioxus*		*Schizaster–chiajei*	*Chione±*	*Lima hians*	
Locality	601	606	613	Figarola	Sv. Ivan	Banjole	617
Depth	30 m	15 m	9 m	20.5 m	44–51 m	35 m	35 m
Total							
Species	61	108	137	73	60	54	169
Wet biomass	48.6	338.7	2,212.6	6,353.4	4,179.5	106.1	2,375.0
Endobenthos							
% Species	62.1	37.0	28.5	2.7	3.3	0	14.8
% Wet biomass	95.9	78.4	1.2	trace	trace	0	0.2
Mobile epibenthos							
% Species	37.7	51.9	54.7	38.4	38.4	24.1	45.0
% Wet biomass	3.9	21.1	98.6	53.1	9.0	2.3	18.0
Sedentary epibenthos							
% Species	9.8	11.1	16.8	58.9	58.3	75.9	40.2
% Wet biomass	0.2	0.5	0.2	46.9	91.0	97.7	81.8

Data for localities 601–617 are compiled from Zavodnik 1971. Biomass is total grams of material examined. Note that total number of species for Figarola, Sv. Ivan, and Banjole is less than number of species locally in the community (see text).

year's total species richness and sum of biomass is used. The other three sites were reported by McKinney (2003), one of which (Banjole) was studied from 25 × 25 cm quadrats collected by a diver, while the other two were sampled with a Musular dredge, which cuts into the sediment no better than the Charcot dredge.

LIM CHANNEL (ZAVODNIK 601)

Lim Channel (fig. 7.9: arrowed locality no. 1) is a deep (35 m), narrow, nearly linear, structurally controlled, flooded valley that intersects the coast north of Rovinj. A few local freshwater springs flow into it, but the salinity is little different from that of the adjacent open water. The channel is floored by muds that are residual from solution of limestones of the adjacent karst plateau. Station 601 was located about midway along the marine portion of Lim Channel, where approximately half the sediment was clay and half silt (Zavodnik 1971). Biomass of the dredged samples consisted largely of a carnivorous gastropod, an irregular echinoid, and a polychaete, followed by a holothurian and diverse bivalves. (Organisms in this and sub-sequent lists are given in order of decreasing mass.) Almost all of the animals in the community are endobenthic (table 7.6), with less than 4% mobile epibenthos and a trivial amount of sedentary suspension-feeding epibenthos. The station was located within an area that was densely sampled by Vatova (1935), and the proportions of endobenthos and epibenthos were only slightly different from Vatova's closest grab samples.

Valdibora Bay (Zavodnik 606)

Valdibora Bay is an open coastal embayment immediately north of Rovinj (fig. 7.9: no. 2). It is floored by fine sediments derived from the adjacent karstic terrain, although they are slightly coarser than the clay-rich muds of Lim Channel. Less than 20 percent of the sediment is clay, about half is silt, and the remainder is very fine to fine sand with a small amount of coarser sand (Zavodnik 1971). Two-thirds of the biomass at Zavodnik's site 606 in Valdibora Bay consisted of a single species of carnivorous gastropod, followed by epibenthic gastropods, asteroids, and crustaceans, plus smaller masses of ophiuroids, bivalves, and scaphopods. The most abundant megaguilds in this community are endobenthic/endobenthic-probing carnivores (gastropods, asteroids, scaphopods), followed by epibenthic grazers (hermit crabs), endobenthic suspension-feeders (ophiuroids), and endobenthic deposit[feeders (bivalves). Endobenthos was moderately underrepresented in the dredged samples (table 7.6: about 80 percent) compared to Vatova's nearby grab sample (99 percent).

Dvije Sestrice Islands (Zavodnik 613)

At 9 m depth Dvije Sestrice was the shallowest of the seven sites (fig. 7.9: no. 3), and it was partially covered by marine grasses. Waves and currents at the shallow, exposed site have winnowed the sediments of almost all fines; it is about 70 percent medium sand with admixtures of fine sand and of coarse sand to gravel (Zavodnik 1971). Over 90 percent of the biomass consisted of holothurians, followed distantly by regular echinoids, gastropods, crustaceans, bivalves, starfish, and polychaetes. The holothurian was an epibenthic detritus-/deposit-feeder, and other well-represented megaguilds included epibenthic grazers, epibenthic carnivores, and endobenthic suspension-feeders. These dredge results are rather different from Vatova's nearby grab sample, which had 92 percent endobenthos comprised of diverse deposit-feeding polychaetes, in contrast with the 28 percent endobenthos (table 7.6) comprised largely of bivalves in Zavodnik's dredged samples. Zavodnik (1971) noted the extreme patchiness of bottom conditions, including patchiness of distribution of marine grasses in the area encountered by the dredge, which may account at least in part for the rather extreme disparity in results.

FIGAROLA

Biomass of the dredged samples off Figarola Island (fig. 7.9: no. 4) was dominated by epibenthic holothurians, followed by ascidians, then by much smaller biomass of rhodophytes, poriferans, bryozoans (predominantly encrusting species) and others. The most abundant megaguild in this community is mobile, surficial detritus-deposit-feeders (holothurians) followed by attached low suspension-feeders (ascidians, poriferans, bryozoans). Although only a trace volume of endobenthos was collected in the surface dredge at Figarola (table 7.6), the closest grab sample taken by Vatova (1935) contained 9.7 percent endobenthos (largely a deposit-feeding bivalve), and the remaining 90.3 percent was epibenthic sedentary suspension-feeders (largely ascidians).

BANJOLE

Banjole Island is ringed by several local communities of benthos on the locally 35 m deep sedimentary plain (fig. 7.9: no. 5), as indicated by Seneš (1988c; the "target" southwest of Rovinj from which four transects radiate in fig. 7.14), though his several transects from the island did not cross the region that begins a few meters west of the rock cliff base of Banjole, referred to as a *Cellaria* meadow by McKinney and Jaklin (2000). The meadow is approximately 80 percent covered by a dense turf of erect bryozoans, largely colonies of *Cellaria fistulosa*, which grows as rigidly calcified, articulated branch segments about 0.5 mm in diameter and a few mm in length, making bushy colonies 10–15 cm high (fig. 7.13A). Biomass of the diver-collected samples from the *Cellaria* meadow was overwhelmingly dominated by erect bryozoans (*Cellaria* plus numerous others), with small proportions of poriferans, encrusting bryozoans, ascidians, polychaetes, arthropods, ophiuroids, and gastropods. Holothurians can be observed in the *Cellaria* meadow from which the Banjole samples were taken but are much less abundant than in the adjacent areas where the seafloor is less densely covered by bushy bryozoan colonies.

Benthos in samples from the Banjole *Cellaria* meadow consists almost entirely of attached suspension-feeders (table 7.6) and appears to be an accurate representation of the structure of the local benthic community (fig. 7.15). The samples were collected by a diver who cut around complete sections of the turf, getting everything down into the sediment, and cores of sediment within the meadow have no indication of burrow-generated churning.

Vatova (1935) took a grab sample near Banjole, but on the north side. The endobenthos of that sample comprised only 15.5 percent of the biomass, mobile epibenthos comprised 20.7 percent, and sedentary epibenthos 63.8 percent. The largest biomass in Vatova's sample consisted of the free-lying epibenthic bivalve *Arca noae*, followed by ascidians, the endobenthic bivalve *Chione*, and the epibenthic predatory gastropod *Murex*. Epibenthic

FIGURE 7.15 The *Cellaria* meadow west of Banjole Island.
A) Lateral shot across a part of the meadow almost exclusively made of *Cellaria*, with marine snow suspended in water. B) oblique view onto surface of meadow with *Cellaria* supplemented by *Pentapora fascialis* and other erect calcified bryozoans; note the crinoid, *Antedon mediterranea*, perched on top of the bryozoans left of center (indicated by arrows).

crustaceans and holothurians, plus diverse endobenthos (sipunculids, ophiuroids, polychaetes, and diverse small bivalves) comprised the remaining 5 percent of the biomass. Vatova (1935) assigned this grab sample to the *Chione verrucosa* zoocoenosis, but on the basis of diver transects through the area Seneš (1988c) considered it to be characterized by epibenthos and assigned it to his *Hippodiplosia foliacea* (= *Pentapora fascialis*) facies. *P. fascialis* is a robust, erect, rigidly calcified bryozoan that grows as a proliferation of bifoliate sheets (fig. 7.13D). *P. fascialis* was not mentioned by Vatova (1935) for the Rovinj area, though it long has been known to grow prolifically in the area and to be the basis for a "coraligene" community there (Nikolić 1959).

Sv. Ivan

A dredged area west of Sv. Ivan Island at 44–61 m depth (fig. 7.9: no. 6) is the deepest of the seven localities. Biomass of the samples dredged from Sv. Ivan was overwhelmingly dominated by erect bryozoans and secondarily by poriferans and ascidians, followed by smaller biomasses of holothurians, asteroids, polychaetes, ophiuroids, and gastropods. The high biomass of the erect bryozoans, poriferans, and ascidians make this the second-most sedentary epibenthos–rich of the localities characterized (table 7.6). As at Banjole, *Cellaria fistulosa* is the most abundant bryozoan. Young colonies of *Pentapora fascialis* constituted the next-highest biomass. Vatova's map indicates a grab sample was taken in the same vicinity, but it belongs to the set for which data were not given in any of his published papers.

I first saw the benthos west of Sv. Ivan in June 1988, and my wide-eyed field notes read: "~60 m deep; about 15 minute dredge, yielding almost a cubic meter of *Pentapora foliacea* (*P.fascialis*) colonies, secondarily *Cellaria*, some shells and cobbles. *Pentapora* and *Cellaria* inhabited by thousands of filter-feeding ophiuroids, a few 10s of *Antedon mediterranea* [a comatulid crinoid]." Unfortunately, only specimens for taxonomic determination were kept; no bulk sample was taken for determining diversity and biomass. Reports from divers collecting bryozoans from a bit shallower water (30–50 m) closer to the island reported that large *P. fascialis* colonies (fig. 7.16) were abundant, growing from isolated small hard substrata or attached to the tough core of dead gorgonian corals above a generally sand bottom.

The next time I visited the locality was early autumn 1990, and dredging in the same area produced only a couple of liters of dead, fragmented material. The area had experienced severe anoxia in autumn 1989, which had eliminated the benthos over a 1,200 km^2 area off the Istrian coast (Jaklin and Zahtila 1990), including the deeper waters near Sv. Ivan. The samples characterized in table 7.6 were taken in summer and autumn 1997. By that time, eight years following the anoxic event, the bryozoan fauna was reestablished as constituting well over half of the epibenthic biomass (McKinney 2003),

FIGURE 7.16 A single 30 cm–high colony of the bryozoan *Pentapora fascialis* growing on a flexible gorgonian coral, part of which projects as a white branch above bryozoan; the gorgonian is attached to the rock at the bottom of the photograph. Collected by hand June 1988 from about 50 m depth west of Sv. Ivan Island.

largely *Cellaria* spp. followed by *Pentapora fascialis*. *P. fascialis* colonies were still less than a third the height of the larger colonies that had been present in 1988. In 1997 ophiuroids, especially *Ophiothrix quinquemaculata* that in 1988 had so prolifically inhabited the interstices of the convoluted sheets of *P. fascialis*, were conspicuous by their absence. Only juveniles were found, and they comprised only 0.2 percent of the wet biomass. It is unclear whether their slow recovery in the area was due to slow influx of larvae or to inadequacy of the small *P. fascialis* colonies to function as surrogate "stems" from which they could feed on suspended matter well above the sediment–water interface.

ZAVODNIK 617

Zavodnik's station 617 is in open water a few km west of Sv. Ivan (fig. 7.9: no. 7), within the offshore sand belt that extends along the Istrian coast. Approximately 50 percent of the sediment at the station was composed of very fine to fine sand, with roughly equal admixtures of mud and medium sand to gravel, including rhodoliths (Zavodnik 1971). Over half the biomass of the dredged samples consisted of ascidians, followed by abundant asteroids, ophiuroids, bivalves, sponges, and holothurians, with lesser proportions of gastropods, polychaetes, and bryozoans; diverse other groups occurred in minor amounts. There was a preponderance (table 7.6) of sedentary epibenthic suspension-feeders (ascidians, bivalves, sponges, bryozoans), followed by epibenthic carnivores (asteroids, diverse but small biomass of crustaceans), epibenthic mobile suspension- (ophiuroids) and deposit-feeders (holothurians, ophiuroids), and a small volume of endobenthos (bivalves, diverse polychaetes). The three grab samples taken by Vatova (1935) that surround this station average the same low proportion of endobenthic biomass and close to the same proportion of attached sedentary epibenthic suspension-feeders.

BRYOZOANS

Extremely abundant calcified bryozoans with diverse erect and encrusting growth habits constitute one of the most striking features of the epibenthos in some communities of the flat soft-sediment seafloor in the northeastern Adriatic. The species present in the northern Adriatic are a subset of the over 300 known to live in the Mediterranean (Zabala and Maluquer 1988) and are even reduced from the number likely to be found farther south in the Adriatic (Novosel and Pošar-Domac 2001). Usually the rigidly erect calcified species in the Adriatic are characteristic of rock outcrops rather than soft-sediment substrates. The erect bryozoans in the soft-sediment communities of the northeastern Adriatic are attached to carbonate shells and skeletons, small stones, and to organic structures, e.g., the leathery thecae of the ascidian *Microcosmus* and to erect gorgonian corals. This is the same range of substratum types (mineralized skeletal material, semifirm organic substrata, rock debris) required for recruitment of erect Paleozoic bryozoans on shallow, soft-sediment seafloors.

Calcified erect bryozoans were more common than encrusters on soft-sediment bottoms of tropical and temperate shallow seas during the Paleozoic (Jackson and McKinney 1990). As the Paleozoic bryozoan fauna became established during the Ordovician and as the post-Paleozoic bryozoan fauna became established during the Jurassic, there were roughly equal numbers of species with erect growth habits and with encrusting growth habits. However, the trajectories of the two growth-habit groups were different during the two time intervals. By the Permian period fewer than 10 percent of shelf-depth calcified bryozoans were encrusters, but by the Late Cenozoic about 70 percent were encrusters.

Free-lying mound-shaped and unattached sheet-shaped bryozoans also were common from Ordovician through Early Carboniferous, declining through Late Carboniferous and Permian (McKinney and Jackson 1989).

Most free-living post-Paleozoic bryozoans have been cap-shaped forms, many having specialized zooids with long, rigid cuticular structures that enable them to right themselves if turned over, to crawl, or even to burrow upward through sediment if they are buried (Cook and Chimonides 1978). Free-lying mound-shaped bryozoans are uncommon in modern bryozoan faunas (McKinney and Taylor 2003), and free-lying unattached sheet-shaped colonies are essentially unknown except in the northeastern Adriatic (McKinney and Jaklin 1993).

The conspicuous erect calcified and free-lying bryozoans, along with suspension-feeding ophiuroids, give several benthic communities off the Istrian coast an eerily Paleozoic structural appearance. Some of the more notable of these bryozoans are discussed below, with information on local distribution in the vicinity of Rovinj taken from Hayward and McKinney 2002.

Pentapora fascialis

Pentapora fascialis grows as bifoliate sheets and branches of variable width that frequently divide and may anastomose (fig. 7.13D). Older colonies growing in open areas offshore of Rovinj generate massive, crinkly heads of complexly twisted sheets and branches that may be closely to loosely spaced. Most colonies collected by dredging appear to have a chaotic intermixture of moderately broad sheets and branches of variable width.

P. fascialis colonies add 2–3.6 cm growth to the edges of the bifoliate sheets each year (Bullimore 1987; Pazold et al. 1989; Cocito and Ferdeghini 1998), allowing an estimate of 8–15 years age for the colony in figure 7.16. Much larger colonies occur adjacent to freshwater springs (vruljas) along the Croatian coast south of Istria, with colonies averaging 66 cm in diameter immediately adjacent to the vruljas and some reaching 150 cm (Cocito et al. 2004; Novosel et al. 2005). The extraordinary size of these colonies may be due to the introduction of nutrients via the fresh water and consequent microbial growth providing a rich food source. At the springs stenohaline animals such as cyclostome bryozoans fouling proximal regions of *P. fascialis* colonies and abundant crinoids (*Antedon mediterranea*) perched on them (Novosel et al. 2005) suggest that although adjacent to the outflowing fresh water the colonies are seldom bathed by it.

Similarly large sizes of *P. fascialis* are reached elsewhere in the Mediterranean, best documented on rocky substrata in the Ligurian Sea (Cocito et al. 1998a, b; Cocito and Ferdeghini 2001). As is the norm for large colonies of erect bryozoans, Cocito and colleagues noted that proximal portions of branch systems typically were dead and fouled by sediment and diverse encrusting organisms, and some entire branched sheet systems were dead all the way to the outer tips. Following a vigorous storm that hit the Ligurian study site, growth of the few survivors, plus recruitment of new colonies and fusion of contacting colonies, rebuilt the same dense large-colony cover within three and a half years (Cocito et al. 1998b). This is much more rapid recovery than seen at Sv. Ivan near Rovinj following the 1989 anoxic event (see above).

P. fascialis is one of the largest of the living bryozoans and is a vigorous producer of carbonate. In the Ligurian Sea standing stock ranges up to almost 2,500 g m^{-2} and annual production up to 1,200 g m^{-2} y^{-1} (Cocito and Ferdeghini 2001), and standing stock at the vruljas of the Croatian coast reaches a remarkable 5,800 g m^{-2} (Cocito et al. 2004). All this skeletal mass is not matched by a huge dry mass of organic matter, which is only about 3 percent or so of the skeletal mass (Cocito and Ferdeghini 2001).

This species exhibits ecophenotypic plasticity in regions of strong unidirectional currents (Cocito and Ferdeghini 2000). The strongest unidirectional currents at the Ligurian Sea study site can reach 0.9 m sec^{-1}, producing colonies of sheets oriented perpendicular to flow. Colonies located on promontories so that they experienced the unidirectional current on the upstream side but turbulent drag on the downstream side were zoned into sheets oriented perpendicular to the oncoming current on the upstream side versus more chaotically oriented bifurcating and irregularly coalescing branches on the downstream side, but without any change in colony density between the two colony morphologies. *P. fascialis* in the Rovinj area experience reversing tidal flow, and side-to-side zonation in sheets versus branches is not conspicuous. Some colonies offshore of Rovinj apparently live in spots where flow overall must almost lack any unidirectional aspect because they are made almost entirely of branches rather than sheets (Nikolić 1959).

P. fascialis is widely distributed and common offshore of Rovinj. It can be found on rock walls and boulders from 10 to 50 m deep and attached to shells and skeletal substrata on the soft sediment floor from 20 to 61 m deep. Typically, only the outer 2–4 cm of the sheets are covered by living tissue, and proximal portions of colonies provide a semicryptic protected environment for abundant and diverse encrusting and erect bryozoan species plus other attached organisms.

CELLARIA FISTULOSA AND C. SALICORNIOIDES

Cellaria fistulosa and *C. salicornioides* are erect bryozoans with small, radially symmetrical, rigidly calcified branches (fig. 7.13A). The colony as a whole, however, is not rigid. A constriction develops in the calcified skeleton as bifurcation points are growing, beyond which the skeleton expands to its normal branch diameter. The calcite of the narrow portion is then resorbed or possibly breaks, with the end result that a flexible, stress-absorbing cuticular pad connects the daughter branches with the parent. Young colonies of *C. fistulosa* are recruited onto hard substrata or firm organic substrata, and as the branch systems proliferate, specialized zooids extend cuticular tubes more-or-less proximally to come into contact with older branches or any hard or firm object, to which they then become attached. Off the Atlantic coast of France, a congeneric species grows at rates of approximately 4 cm yr^{-1} (Bader 2000), generating approximately three successive branch segments within the year. New branches of *Cellaria* usually are added by distal bifurcation, but lateral branching also occurs from older branch segments deep within the colonies (Bader 2000; McKinney and Jaklin 2000).

The two *Cellaria* species offshore of Rovinj are even more widely distributed and common than is *Pentapora fascialis*. They range at least from 5 to 30 m deep on rock walls and boulders and also recruit onto shells, skeletal substrata, and *Microcosmus* on the soft sediment floor 22–61 m deep. In the *Cellaria* meadow (fig. 7.15), dense continuous growths extend up to at least 50 cm wide but commonly no more than 15 cm deep; it is impossible to determine whether these individual turf units are a single colony or are composed of multiple intergrown colonies. They have spread well beyond the original hard substratum, apparently growing laterally over the substratum as a bulging lateral front of branch tips, with the proximal portions of the colony or colonies accumulating detritus from in situ skeletal production of the *Cellaria* and its epibionts and from baffling of silt from turbid water during tidal flow.

For the larger segments of *Cellaria* turf, it is impossible to find the point(s) of origin. The turf typically shows no sign of senescence, with the lateral branching deep within the turf able to renew any areas of the surface that become damaged. It is host to over 80 species of attached epibionts, each of which has a characteristic depth range within the turf (McKinney and Jaklin 2000).

RETEPORELLA GRIMALDII

Reteporella grimaldii, for many years identified as *Sertella septentrionalis*, grows as narrow, sinuous, anastomosed, unilaminar branches that generate fenestrate colonies (fig. 7.13E). Just as *Pentapora fascialis* and *Cellaria* spp., *R. grimaldii* colonies can grow on a diversity of hard and organic substrata, even live portions of gorgonian sea whips (Hass 1948).

Hass (1948) described extensive morphological variation in *R. grimaldii*. Colonies vary in branch robustness, number of zooid rows per branch, distance between successive points of anastomosis, tightness of longitudinal folding of local colony surfaces, intensity of colony convolution, and overall colony openness. No formal analysis has been made within the Rovinj area, but a qualitative assessment of these variables suggests that the variable colony morphologies are ecophenotypic, with greater exposure to current and wave motion resulting in greater branch robustness, a larger number of rows of zo-oids per branch, more closely spaced points of anastomosis, and more com-plexly folded and convoluted fenestrated sheets, generating denser colonies.

R. grimaldii is widespread and moderately common offshore of Rovinj. It grows from 3 m to at least 30 m deep on rock surfaces and boulders, tend-ing to occur on more cryptic surfaces at the shallower depths, and from 22 to 61 m deep on local hard and organic substrata on the soft-sediment floor. Colonies in protected areas under boulders can grow to almost 1 m in diameter (Davor Medaković, personal communication) but seldom exceed 10 cm diameter in exposed areas.

RHYNCHOZOON REVELATUS

Rhynchozoon revelatus is an encrusting species that both spreads laterally across the substratum and adds new zooids frontally so that it thickens as well. It recruits onto hard substrata and onto thecae of the ascidian *Micro-cosmus*. Because of vigorous frontal budding, colonies are highly plastic morphologically and can respond to environmental stimuli to form rela-tively expansive thin colonies to more laterally constrained, irregularly nodular colonies. The frontal budding allows growth of small mounds, called monticules, spaced a few millimeters from one another as a normal part of the colony morphology. Monticules were typical phenomena of Pa-leozoic stenolaemate bryozoans that had broad colony surfaces; they are in-terpreted as focal points from which water is jetted away from the colony after it has been drawn to the general colony surface and filtered through the feeding structures. The monticules of *R. revelatus* clearly perform that function (fig. 7.13F).

R. revelatus was not found on rock and boulder surfaces off the Istrian coast but occurred on *Microcosmus* and hard substrata on the soft-sediment plain from 20 to 45 m deep. It commonly grows laterally beyond its substra-tum, which can result in free-lying domes (fig. 7.13F) reaching diameters up to 4 cm and thickness over 1 cm, very much the size and shape of numerous Paleozoic trepostome and cystoporate mound-shaped bryozoans.

CALPENSIA NOBILIS

Calpensia nobilis is a very common species in the Rovinj area that grows rap-idly as a single-layered sheet upon a huge diversity of natural and artificial substrata. The expanding sheets can overgrow themselves to make multilayered colonies, and small substrata can be completely engulfed as a core of a rolling mass expanding sheet-by-sheet. Expansive free-lying sheets of *C. nobilis* (fig. 7.17) were discovered in early December 1989 by divers tending bottom-tethered instruments about 20 km southwest of the southern tip of Istria (McKinney and Jaklin 1993). Colonies were littered 1–2 m^{-2} across the 45 m–deep, muddy sand bottom in the vicinity of the instruments. The site was revisited September 1990 specifically to collect additional samples of *C. nobilis*, but despite concentrated search by divers in the same area and prolonged dredging around the instrument site, only corroded fragments of colonies about 1 cm diameter could be located.

A similar bloom was discovered early October 1990 about 40 km south of the southern tip of Istria (see X on fig. 7.1), again in the vicinity of tethered instruments. Numerous colonies up to 14 cm across were collected from the 49 m–deep muddy sand bottom and were taken live to a laboratory for study. The colonies were extraordinarily fragile and had a specific

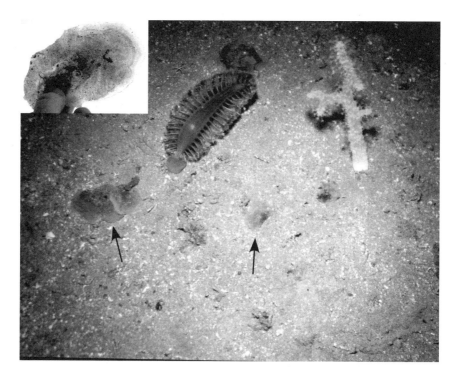

FIGURE 7.17 Two free-lying thin colonies of *Calpensia nobilis* adjacent to the sea pen *Pteroeides spinosum* and a dendroid octocorallian, possibly *Alcyonium acaule*, 49 m deep, October 1990 about 40 km south of the southern tip of Istria. Inset shows a *C. nobilis* colony from the same bloom, about the size of the smaller of the two arrowed colonies.

In situ photograph courtesy of A. Jaklin.

gravity estimated to be no more than 1.16, little more than that of the ambient seawater (1.03); they lay as feathers on the seafloor.

Free-living *Calpensia nobilis* were first discovered off southern Istria about a month after a major anoxic event off Istria in mid November 1989, but while the site was deep enough to be involved in the anoxic event, it was on the southern fringes of the area known to be affected (McKinney and Jaklin 1993). No such *C. nobilis* colonies were present in August 1989 during the previous visit by divers to tend instruments, indicating that the colonies had grown during the four-month interval. The maximum growth rate of *C. nobilis* is not known, but related weedy, calcified anascan bryozoans can extend their margins almost 1 cm day^{-1} (Udhayakumar and Karande 1989). If the *C. nobilis* colonies grew at even half that rate around their perimeters, they could have reached the observed size within the three weeks following the anoxic event.

The stimulus for the two blooms is unknown. Similar blooms did not occur where regularly monitored benthic stations were located within the 1989 anoxic region. Although *Calpensia nobilis* comprised an estimated 90 percent of benthos dredged from the site of the first bloom, some fecal strings of sediment-feeding benthos along with mature *Ophiothrix quinquemaculata* and other epibenthos were present along with them. Nonetheless, the opportunistic invasion of *C. nobilis* colonies that spread across the sediment substrate appears to have been related to temporarily reduced bioturbation. The growth habit had not been seen prior to the increase in size and intensity of anoxic events in the late 1980s, even though the eastern northern Adriatic has relatively low background bioturbation rates as indicated by low levels of disturbance of sediment lamination (Puškarić et al. 1990). Recently a lower-density population has been found offshore of Rovinj in the outer silt belt (Andrej Jaklin, personal communication). The near-Rovinj population is longer-lived than the two more southerly blooms, existing for over a year, and its presence has no obvious relation to an anoxic event.

ISOLATED WESTERN HARD SUBSTRATA

Hard substrata of various origins are found in many places in the Adriatic (Newton and Stefanon 1982). The most extensive are the submerged parts of coastal cliffs and islands, and the totally submerged carbonate peaks off the Balkan coast that are the result of drowned karst topography. Yet there are some local hard substrata farther west in the Adriatic, in addition to human constructions in the sea. Seeps of gas, both methane and CO_2, derived from decomposing organic material in the underlying sediments, locally result in carbonate cementation of relict sands. Some seep-influenced sands are sufficiently indurated to stand as pinnacles if the surrounding less coherent sands have been eroded away. Such pinnacles are scattered throughout the northern Adriatic and are not uncommon, but much information on their exact localities is withheld by scientists in the hopes of limiting overfishing by sports divers of both the encrusting organisms and the associated benthic fish (Conti et al. 2002).

Benthos on Western Hard Substrata

Communities of attached and mobile benthos on rocky substrata of the eastern coast have been described by Vatova (1935, 1949), Gamulin-Brida (1974), and others. This book is focused on the benthos of sedimentary substrata, so these prolific communities are not explored here. However, hard substrata of any size farther west are of interest because they give insight into the range of organisms that potentially are present but generally uncommon or absent except along the eastern coast.

Drowned beachrock of ancient strandlines, and low biogenic structures built by calcareous algae, seruplids, bryozoans, and sponges are found along the boundary between modern mud along the shore off the Venice lagoon and residual sands farther offshore (Colantoni et al. 1979; Newton and Stefanon 1982). Extensive growth of calcareous algae and anthozoans on drowned Holocene beachrock has created numerous reeflike structures at about 20 m depth off the south end of the lagoon of Venice (Colantoni et al. 1979; Newton and Stefanon 1975). The structures have low slopes but are asymmetrical, with somewhat steeper slopes facing the open sea.

Epibenthic macrofauna are diverse and dominated by suspension-feeders on these structures, as documented by Gabriele et al. (1999). They compared the epibenthos of a near-shore and a farther offshore beachrock community and a near-shore sunken ship at the same depth as the near-shore beachrock. Taxa recorded among the three stations included ascidians (23 spp.), bivalves (13), gastropods (12), sponges (11), annelids (6), anthozoans (5), holothurians (2), ophiuroids (2), sipunculids (2), a bryozoan, and a hydrozoan. The total number of species encountered on the near-shore beachrock was only 40 percent that of the other two stations. Differences in relative abundance of taxa among stations studied by Gabriele et al. appear to be due to sedimentation rate and wave exposure, both of which influence turbidity, and slope of the substratum, which interacts with turbidity to determine sediment accumulation rate. The beachrock had shallower-sloped surfaces than did the ship. Turbidity was high, producing less than 2 m visibility at the near-shore beachrock and at the ship but was lower, allowing visibility about 6 m, at the offshore beachrock.

Off the lagoon of Venice, biomass (measured as ash-free dry weight) of both ascidians and sponges was high on the offshore beachrock, but fell off by 50 percent for ascidians and by over 90 percent for sponges at the near-shore beachrock (Gabriele et al. 1999). Attached bivalves were virtually absent on the natural outcrops but were abundant (especially *Ostrea edulis*) on the ship. Biomass was higher on the ship than at the other two stations because of the abundance of bivalves. Between the two natural substrata, there was higher biomass per unit area at the offshore site in less-turbid water, where there was almost equal abundance of ascidians and sponges. Apparently the more intense sediment-smothering of the near-shore beachrock, where turbidity was higher, was the primary cause for suppressing biomass and species richness of attached benthos there.

In addition to the larger substrata, living and relict algal rhodoliths, even coralline-based patch reefs, may be found in regions of low sedimentation (Bressan and Nichetto 1994; Aiello et al. 1995). Algal rhodoliths occur between the Po Delta and the Gulf of Trieste, increasing in number offshore and reaching maximum density about midway between the delta and the gulf. Heavily encrusted shells, especially shells inhabited by hermit crabs, and shell complexes are widely distributed and common in the northern half of the Gulf of Trieste (Zuschin and Piller 1994; Zuschin and Perversler 1996).

Larval Supply

Larvae of epibenthic sedentary suspension-feeders are widely available in the western region of the northern Adriatic. They settle on hard substrata in lagoons (Bianchi and Morri 1996; Cornello and Manzoni 1999), on natural and artificial hard substrata north of the Po Delta as indicated above, and also on hard substrata south of the Po Delta. Forty species of such organisms settled on two gas platforms in eutrophic, plankton-rich waters offshore of Ravenna (Relini et al. 1998). Sedentary suspension-feeders on the platforms and associated study panels included sponges, hydroids, anthozoans, polychaetes, barnacles, bivalves, bryozoans, and ascidians, plus mobile suspension-feeding ophiuroids.

The occurrence of diverse epibenthos, many of which are sedentary suspension-feeders, on the locally available hard substrata off the Italian coast of the northern Adriatic suggests that their scarcity across the sedimentary seafloor of the region is not due to larval absence. Circulation in the northernmost Adriatic is cyclonic (figs. 4.4, 4.6), which would be the path traveled by planktonic larvae. The isolated natural hard substrata along the path that sweeps by Istria and then follows the Italian coast are sufficiently abundant and closely spaced along the entire coast to permit transfer of even short-lived larval from the epibenthos-rich area off of Istria.

In fact, a moderately diverse suite of bryozoans, a subset of the taxa that are so abundant offshore of Istria, occurs in deeper water off the Italian coast from about 50 km south of the Po Delta to Ancona (Poluzzi 1979). The fauna is depauperate and consists largely of erect articulate colonies under the seaward margin of the Po plume but diversifies rapidly in taxa and growth habits farther offshore on relict Pleistocene sands. Although the bryozoans are not known to be as abundant as off the Istrian coast, even colonies of rigidly erect robust species such as *Pentapora fascialis*, *Smittina cervicornis*, *Schizotheca serratimargo*, and the fenestrate species *Reteporella grimaldii* occur at 40 m and greater depths (Poluzzi 1979). Absence of a richly developed epibenthic fauna on the soft-sediment seafloor along the Italian coast is due to something other than absence of larvae for settlement. Possible reasons for this will be explored in chapter 9.

SUMMARY

Endobenthos is ubiquitous across the northern Adriatic, reaching its highest biomass per unit area in the western portion, but also locally abundant in a few eastern onshore areas. In contrast, epibenthos is not ubiquitous but is of low biomass and almost entirely mobile except on local hard substrata across much of the western portion; it is more common and reaches high biomass in the east. Sedentary epibenthos is even more restricted in distribution, reaching its acme within a few kilometers of the Istrian coast and apparently spreading out farther toward the middle of the basin south of Istria. Endobenthic communities of the western regions seem to have greater continuity in distribution than do the communities rich in sedentary epibenthos, which have a much finer-grained texture of distribution along the Istrian coast.

Disruption or destruction of western endobenthic communities, e.g., by anoxia, is followed by rapid repopulation by endobenthic animals that reconstitute a generally similar trophic structure even though different taxa may be involved initially. Eastern epibenthic communities do not recover readily after catastrophic events, requiring at minimum several years or decades before similar communities return; they can be profoundly set back by relatively minor disturbances during the recovery period.

Whether the endobenthos- versus epibenthos-dominated regions can be recognized in the fossil record depends on how faithfully their skeletal remains reflect the original ecological structure. The following chapter examines the potential long-term residues of the Adriatic benthos in order to see how well the taxonomic and megaguild composition might be represented in the fossil record.

CHAPTER EIGHT

FOSSILIZATION POTENTIAL OF NORTHERN ADRIATIC BENTHOS

No organism wholly soft can be preserved. Shells and bones decay and disappear when left on the bottom of the sea, where sediment is not accumulating.
—CHARLES DARWIN, *The Origin of Species*

Darwin was overly pessimistic about the possibility of preservation of soft-bodied organisms. Nevertheless, preservation of an organism's remains as part of the fossil record is a game of chance, with the odds heavily stacked against organisms that live on land and even more so against those that are soft-bodied.

With this in mind, can the benthos of the northern Adriatic Sea be taken at face value as a "window" into ancient ecology? The degree to which life and the local ecological structure of ancient seas can be understood depends upon the fidelity with which the fossil record reflects original life associations. The quality of the fossil record depends upon a host of biological, chemical, physical, and eventually human variables acting from the time that the organisms are alive to the time that their fossil remains are studied. In a normal shallow marine setting, the first filter between living organisms and their probability of eventual fossilization is whether a mineralized skeleton is present. This chapter addresses the potential skeletal residues of the northern Adriatic benthos and how closely they might record both taxonomic diversity and ecological structure of the living communities.

SOME GENERAL ESTIMATES OF PRESERVATION POTENTIAL

It is impossible from an assemblage of only mineralized fossils to determine the species richness of the original life community (or mixed communities) represented. Diverse paleontologists have worked with benthic life assemblages in modern seas to estimate how well they might be represented in the fossil record based on skeletons and apparently taxon-specific traces. Some of the studies have focused only on highly preservable groups such as corals (Edinger et al. 2001) or mollusks (Valentine 1989). Others have examined all the megascopic benthic species in a specific environment, and these are more germane here.

Species richness of potentially preservable coral-reef faunas has been estimated to range from 2 percent (Newell 1959) for the fauna overall to 38 percent for cryptic encrusters (Rasmussen and Brett 1985). Other shallow-water benthic faunas have been determined or estimated to have from about 20 percent to about 40 percent potentially preservable species (e.g., Lawrence 1968; Driscoll and Swanson 1973; Stanton 1976; Schopf 1978; McKinney 1996; Radenbaugh and McKinney 1998), including some northern Adriatic benthic faunas.

Mineral composition, microtexture, and size parameters are all inherent characters of skeletal materials that determine the preservation potential in various physical and chemical environments. A large body of literature has accumulated on preservation potential of skeletal types, and paleontology/paleobiology textbooks generally devote a chapter to the topic. The basic patterns are 1) among the three most common skeletal materials of organisms larger than microplankton, calcium phosphate is most durable chemically at the earth's surface, trigonal $CaCO_3$ (calcite) is next, and orthorhombic $CaCO_3$ (aragonite) is least stable; 2) within each shell composition group, the interplay of crystalline growth habit and intermixed organic matter determines the "toughness" of the shell material; and 3) all else being equal, the larger and more equidimensional the skeletal element, the more durable it will be.

The obvious conclusion from the preceding paragraph is that a large, densely textured calcium phosphate bone has a higher chance of being preserved than a thin, aragonitic bivalve shell with prismatic structure such that the elongate crystallites are perpendicular to the shell surface and are associated with minimal shock-absorbing organic matter. General patterns in the fossil record raise questions about such assumptions, however. The mollusk class Bivalvia, for example, has an abundant fossil record through almost the entire Phanerozoic, and its members variously have aragonitic, mixed, or calcitic skeletons with varying organic content and a range of shell microstructure.

Kidwell (2005) examined each of these three variables for almost 500 bivalve genera and compared each with the temporal distribution (composition of faunas through time and duration of the individual genera as indicators of preservation bias. Preservation bias was expected to result in progressively older faunas with genera characterized by higher proportions of low- to moderate-organic, calcitic shells in which there was smaller crystallite surface area to volume (SAV) relative to younger faunas. SAV is essentially a measure of size and shape of crystallites, with small needle-shaped crystallites having high SAV and large equidimensional crystallites having low SAV.

Of the three characteristics, only crystallite SAV fit the expected trend. Moreover, Kidwell (2005) grouped the bivalves into two contrasting groups based on predicted preservation potential. She compared the duration of ranges for genera that are preservationally "at risk" (aragonitic, high SAV, high organic content, or a combination of these) with the more stable states and found essentially no difference in generic durations of the two groups. The only significant difference in the comparisons was contrary to expectations: the "at risk" group was disproportionately represented among genera with moderate durations of about 20 million years. Despite preservational vicissitudes within local assemblages, Kidwell (2005:917) concluded, "Many skeletonized macroinvertebrate groups in the fossil record have compositions lying within the range spanned by bivalves, and thus there is reason to suspect that differences in skeletal mineralogy and microstructure might have as little net impact on first-order evolutionary patterns as found here among marine bivalves."

Looking more broadly, Behrensmeyer et al. (2005) examined skeletal durability versus frequency of occurrence of the 150 most widely encountered genera of the common fossil groups Brachiopoda, Bivalvia, and Gastropoda. Each genus was scored for character states that affect durability of shells: overall size, relative thickness of the shell, skeletal reinforcements (folds, ribs, spines), organic content of the shell, and mineralogy. The latter two characters were scored only for bivalves and gastropods, both of which may have either calcitic or aragonitic shells (mixed calcite-aragonite shells were omitted) with variable organic content. Frequency of occurrence was based on the number of named sedimentary rock formations in which the genus is known to occur.

Surprisingly, shell durability is essentially not correlated with commonness in the fossil record for these three major groups. Of 29 comparisons between character states in the three groups and the measures of shell durability, 23 showed no significant statistical correlation between shell durability and commonness, three were significantly correlated positively, and three were significantly correlated negatively, i.e., the less robust state was the more common (Behrensmeyer et al. 2005). These findings, which are contrary to laboratory and thought experiments, could be the result of several mitigating factors, such as initial abundance of the living animals, animals with more delicate skeletons living preferentially in less destructive environments, etc. Whatever the reason for the results of the Behrensmeyer et al. study, information about some of the mineralized benthos in the northern Adriatic suggests that they too are broadly preservable.

The intertidal zone of the coast of the Po Delta and the marsh margins of Commachio, a large estuary south of the delta, is comprised of three general types of substrates: hard surfaces, sands, and muds. Approximately 37 percent of the benthic species of the sands and muds have indurated, well-mineralized skeletons, and only a slightly lower proportion (32 percent) of those on hard substrata have a high preservation potential (Poluzzi and Taviani 1988). Approximately half the remaining species could possibly be represented in the fossil record by less durable structures such as spicules and spongy mixtures of organic material and calcite. Poluzzi and Taviani (1988) reported that all but 2 of the 40 species with high preservation potential, and 8 (questionably more) of the 44 species with mediocre preservation potential, had previously been found as fossils.

Fifty-nine genera of bryozoans with mineralized skeletons were recorded by Hayward and McKinney (2002) off the Istrian coast of the northern Adriatic. Some are characterized by colonies that are composed of delicate and lightly calcified branches while others have large, heavily calcified colonies composed of robust branches or forming a single mass several centimeters in diameter. Yet all 59 genera have extensive fossil records in Neogene circum-Mediterranean sedimentary rocks. Two faunas in Pliocene sedimentary units deposited from the Adriatic Sea together contain 83 percent of the genera described by Hayward and McKinney (2002) within even more diverse bryozoan faunas. One of the faunas (Poluzzi 1975) is located well to the west in the Po plain, near Piacenza, and occurs in Pliocene muds generally similar to those being deposited near the Po Delta at present; it includes 61 percent of the mineralized bryozoan genera found at present offshore of Istria. The other (Poluzzi et al. 1988) is on the edge of the Apennines just west of where they meet the coast near Rimini and includes 65 percent of the mineralized bryozoan genera reported from offshore of Istria. It comprises a clean bryozoan calcarenite containing large volumes of branch segments from articulated colonies.

A systematic survey of the relative durability of the mineralized organisms in the northern Adriatic has not been made. However, the high proportion of Adriatic bivalves and mineralized bryozoans that are known as fossils suggests a high preservation potential even for their more delicate skeletal remains, presuming net sediment accumulation. Therefore, in this chapter I treat all mineralized taxa from the northern Adriatic as having an equal chance of preservation.

Benthic surveys by Vatova (1935, 1949) and Gamulin-Brida (1974), and the geographically more restricted systematic sampling program of Zavodnik (1971), yielded 686 identified benthic and demersal species for the northern Adriatic (table 8.1). This total is slightly more than double that from Vatova's work alone. As indicated in chapter 7, this still does not capture the full benthic diversity within the area. However, these studies were systematically done and were not taxonomically biased other than the individual scientist's differing degrees of ability to identify members of different taxonomic groups. They provide a good base, individually and collectively, for estimation of potential fidelity of preserved remains for taxonomic and ecological interpretation of the ecosystem as a whole.

Aside from brief periods of anoxia in local regions during the transition between late summer and the onset of winter (fig. 5.10), the northern Adriatic Sea is well oxygenated. Therefore, bacterial degradation of soft tissues is ubiquitous, and preservation of organisms that lack mineralized skeletons is unlikely. That eliminates one-third of the higher taxa in the northern Adriatic but relatively few species (table 8.1). Most of the higher taxa that are entirely soft-bodied are represented by few species. The Polychaeta, however, are highly diverse and only 4 among the 143 polychaete species included in table 8.1 secrete mineralized skeletons although several of the remaining species have mineralized jaws.

Approximately two-thirds of the benthic species in the collections included in table 8.1 had mineralized skeletons. Of these, 117 species (17 percent of the total) had skeletons that would not or were unlikely to produce megafossils, including the ossicles of all holothurians, the spicules of mineralized demosponges, and the weakly mineralized, organic-rich skeletons of many crustaceans. Just under half (48 percent) of all the species encountered had indurated skeletons that were potentially preservable as megafossils. Of these, the Bivalvia had the greatest number, followed by Gastropoda, decapod Crustacea, Osteichthyes, Asteroidea (producing numerous disarticulated plates), Echinoidea, and gymnolaemate Bryozoa. Demospongea also were represented by numerous species, although they would be preserved as dispersed spicules. This list of organisms is squarely within the Modern fauna of Sepkoski (1981). In Sepkoski's factor analysis of the post-Paleozoic marine fossil record, Gastropoda and Bivalvia were most important; Osteichthyes, decapod Crustacea, and Echinoidea made up the next-most important group; and gymnolaemate Bryozoa, Demospongea, Chondrichthyes, Polychaeta, Hexactinellida, and Stelleroidea (Asteroidea plus Ophiuroidea) constituted the third group. These three groups of taxa accounted for 93 percent of the Modern axis.

TABLE 8.1 Skeletal Composition and Preservation Potential of Northern Adriatic Megabenthos Encountered in Systematic Sampling Programs of Vatova (1935, 1949), Zavodnik (1971), and Gamulin-Brida (1974)

Major Taxonomic Group	Skeletal Composition	Number of Species Present	Number of Potentially Preserved Species
Demospongea	Organic, silica spicules	39	(29)
Hydrozoa	Organic	3	0
Anthozoa	Organic, calcite spicules, calcite	43	9 + (12)
Platyhelminthes	Organic	3	0
Nemertea	Organic	3	0
Sipuncula	Organic	7	0
Echiurida	Organic	2	0
Polychaeta	Organic, calcite	143	4 + (17[+])
Bryozoa	Organic, mixed organic and calcite, calcite	13	13
Amphineura	Calcite	5	5
Gastropoda	Organic, aragonite	59	53
Bivalvia	Calcite, mixed aragonite and calcite	129	129
Scaphopoda	Calcite	3	3
Cephalopoda	Organic, calcite	5	3
Oligochaeta	Organic	1	0
Phoronida	Organic	1	0
Brachiopoda	Calcium phosphate, calcite	3	3
Crustacea	Organic, mixed organic and calcite, calcite	100	32 + (56)
Chaetognatha	Organic	1	0
Hemichordata	Organic	1	0
Enteropneusta	Organic	5	0
Crinoidea	Calcite	2	2
Asteroidea	Calcite	18	18
Ophiuroidea	Calcite	10	10
Echinoidea	Calcite	13	13
Holothuroidea	Calcite spicules	20	(20)
Ascidia	Organic	25	0
Cephalochordata	Organic	1	1
Osteichthyes	Calcium phosphate	28	28
TOTAL		686	326 + (134[+])

Numbers in parentheses indicate species unlikely to be preserved because they will be preserved as microscopic spicules (Demospongia, octocorallian Anthozoa, Holothuroidea), are easily degraded because of high proportions of organic material intermixed with the mineralized skeleton (many decapod Crustacea), or the only skeletal parts are phosphatic jaw parts (Polychaeta; count from Zahtila 1997).

GENERAL PATTERNS ACROSS THE NORTHERN ADRIATIC

The way in which a fossil assemblage is conceived is dependent on how the components of that assemblage are characterized. Different results are almost always obtained if species are counted as individual units, if the number of specimens of each species is counted, or if a proxy such as skeletal volume or mass is used to estimate biomass. This section describes the relationships of these three measures between living communities and their potential skeletal residues, based on the data published by Vatova.

SPECIES RICHNESS

Within individual grab samples, an average of 49 percent (±19 percent) of the species had megaskeletons (fig. 8.1B). Most of the grab samples had 10 or fewer species, and 30 percent had five or fewer (fig. 8.1A).

Most polychaete species were endobenthic, actively burrowed through the sediment, and had organic skeletons. Their ubiquity, commonly more than one species per sample, offset the widespread endobenthic bivalves, all of which had mineralized skeletons. That, plus the fact that most of the grab samples were dominated by endobenthos, caused the preservation potential of endobenthic species in individual samples (46±24 percent) to be on average identical to the total benthic fauna (fig. 8.1B–C).

Well-calcified epibenthic gastropods and crabs caused the preservation potential of epibenthic species (62±36 percent) to be somewhat higher (fig. 8.1D). The sedentary epibenthos, however, were present in a minority (44 percent) of the grab samples. Species of demosponges and ascidians, both groups having poor preservation potential, were widely encountered, causing the preservation potential of sedentary benthos (22±35 percent) to be conspicuously lower than that of the benthic fauna overall. The high standard deviation relative to the mean preservation potential for sedentary benthos is due to several mineralized bryozoan species occurring along a belt offshore of Istria, as discussed in chapter 7.

SPECIMEN COUNTS

The proportion of individual specimens within a grab sample that potentially leave megaskeletal fossils was more highly variable if there were about 30 or fewer per sample than if there were more. This was the case for all benthic specimens as well as for endobenthic, epibenthic, and sedentary epibenthic (fig. 8.2A–D). However, proportional representation began to stabilize

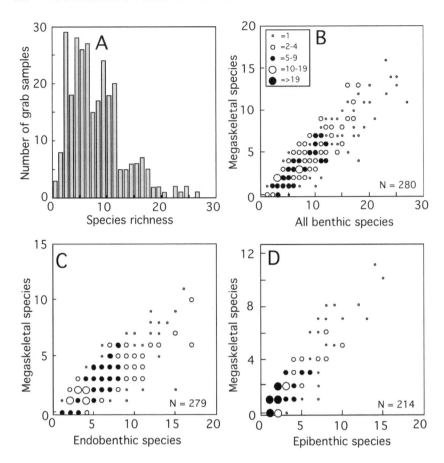

FIGURE 8.1 Number of species per sample and comparison of number of macrobenthic animal species versus number of species with megaskeletons in northern Adriatic
0.2 m² grab samples (Vatova 1935, 1949).
A) Bar graph of number of species per sample. B) All benthic species ($Y = 0.557X - 0.402$).
C) Endobenthic species ($Y = 0.489X - 0.070$). D) Epibenthic species ($Y = 0.621X - 0.007$).
Many samples included few species, resulting in numerous samples plotting on top of one
another in graphs B to D, especially in the lower left of the graphs. The key in graph B indicates the number of samples that each dot represents for each of the three graphs.

above 30 specimens within a sample (fig. 8.2A–C). None of Vatova's grab
samples had more than 30 identified sedentary epibenthic specimens, although offshore of Istria there are many areas where a 0.2 m² grab could encompass many more sedentary epibenthic individuals. A very high proportion
of the sedentary epibenthos offshore of Istria have well-mineralized skeletons
(see below), and their inclusion likely would have produced a pattern similar
to those in figure 8.2A–B.

The extreme variation in preservation potential for low numbers of
endobenthic individuals (fig. 8.2B) reflects the range of endobenthic communities from bivalve-dominated to polychaete-dominated; burrowing

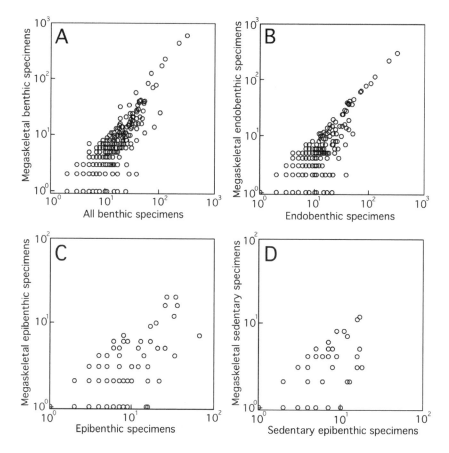

FIGURE 8.2 Comparison of number of individual macrobenthic animal specimens versus number of individuals with megaskeletons in northern Adriatic 0.2 m² grab samples (Vatova 1935, 1949).
A) All benthic specimens (Y = 0.557X – 0.402). B) Endobenthic specimens (Y = 0.935X – 4.969). C) Epibenthic specimens (Y = 0.253X + 0.892). D) Sedentary epibenthic specimens (Y = 0.340X + 0.850).

holothurians although large were seldom so numerous that they contributed appreciably to the total of individuals within a sample. The trend in preservation potential of endobenthos stabilized at nearly 100 percent for samples with over 100 endobenthic individuals because all such samples consisted almost exclusively of dense populations of the suspension-feeding gastropod *Turritella communis*.

The variation in preservation potential for epibenthic individuals (fig. 8.2C) had a more complex origin than that for endobenthic individuals. Some samples were dominated by mobile epibenthos that were either well mineralized (e.g., gastropods, ophiuroids) or were not (holothurians, many crustaceans). Other samples were dominated by sedentary epibenthos (fig. 8.2D) that were either well mineralized (bryozoans) or were not (ascidians, sponges). Yet other samples had roughly equal numbers of mobile and sedentary epibenthic individuals, with numerous taxa determining the ratio of well-mineralized to poorly mineralized to soft-bodied individuals.

BIOMASS AND SKELETAL MASS

Interestingly, wet biomass appears potentially to be represented in the fossil record better than are numbers of individuals. Although the range of proportion of biomass represented by megaskeletons is relatively broad, most data for the benthos as a whole are clustered closely along the 1:1 boundary (fig. 8.3A), and the same is true for endobenthos and epibenthos considered separately (fig. 8.3B–C). The conspicuous deviation away from the 1:1 boundary for endobenthos of small total biomass is due to nonmineralized burrowing polychaetes usually having little biomass compared to bivalves, and grab samples with low endobenthic biomass typically are richer in polychaetes than in bivalves. The greatest deviations from the 1:1 boundary for epibenthos were most commonly caused by holothurians (microskeletal), which on average are larger than other mobile epibenthos, and the numerous deviations for sedentary epibenthos (fig. 8.3D) were due to ascidians (nonmineralized) and/or sponges (nonmineralized or microskeletal), both of which can grow to very large sizes.

INDIVIDUAL SITES

Megaguild structure for some of the northern Adriatic sites at which the macrobenthos in its entirety has been studied by replicate samples is described here. Unfortunately, very few of the studies of biomass have recorded the skeletal or ash residue of the species encountered. Ratios of wet biomass to ash residue for species or related species in Zavodnik (1971) have been used to calculate estimated residues where only wet biomass is known. Other studies give only number of individuals and not biomass, but these too are useful.

ENDOBENTHOS-DOMINATED SITES

PO DELTA
Data on number of individuals per species were given by Parisi et al. (1985) for a site at 8 m depth off the Po Delta (fig. 7.4). The benthic community at this site is comprised of six megaguilds, two epibenthic and the others

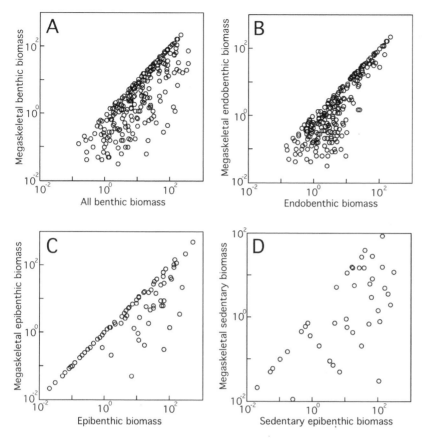

FIGURE 8.3   Comparison of wet biomass of macrobenthic animals versus wet biomass of macrobenthos with megaskeletons in northern Adriatic 0.2 m² grab samples (Vatova 1935, 1949).
A) All benthic biomass ($Y = 0.277X^{1.036}$). B) Endobenthic biomass ($Y = 0.833X - 1.912$).
C) Epibenthic biomass ($Y = 0.659X^{0.810}$). D) Sedentary epibenthic biomass ($Y = 0.337X^{0.595}$).

endobenthic (table 8.2). Only 20 percent of the species of the low-diversity fauna were preservable. Individuals in the two epibenthic megaguilds comprised only a small proportion of the total, and although one of the two megaguilds has no readily preservable species, the epibenthic carnivores survive as a small proportion of the preservable specimens. Although the endobenthic polychaetes all have organic skeletons, their loss eliminates only the endobenthic carnivores, which were represented in the living fauna by less than 3 percent of the total individuals. Endobenthic suspension-feeding and deposit-feeding bivalves represent both megaguilds sufficiently robustly that in the end, the skeletal residues are a fair reflection of the broad ecological structure of the living fauna.

TABLE 8.2   Average Number of Individuals for Megaguilds and Major Taxonomic Groups Present per m^2 at 8 m–Depth Offshore of the Po Delta

MEGAGUILD	TAXONOMIC GROUP	NO. SPECIES	NO. SPECIES IN SKELETAL RESIDUE	NO. INDIVIDUALS	NO. INDIVIDUALS IN SKELETAL RESIDUE
Epibenthic detritus-feeder	Crustacea	2	0	907	0
Epibenthic carnivore	Gastropoda	1	1	129	129
Endobenthic detritus-feeder	Polychaeta	2	0	4,514	0
Endobenthic suspension-feeder	Bivalvia	3	3	3,981	3,981
Endobenthic suspension-feeder	Polychaeta	5	0	2,801	0
Endobenthic deposit-feeder	Bivalvia	1	1	98	98
Endobenthic deposit-feeder	Polychaeta	3	0	584	0
Endobenthic carnivore	Polychaeta	3	0	353	0

Information based on 43 different sets of samples taken in all seasons of the year (Parisi et al. 1985).

OFFSHORE OF RAVENNA

A low-diversity, stress-disrupted benthic fauna occurs at 14–15 m depth offshore of Ravenna. It is almost entirely endobenthic, with only a few non-preservable epibenthic detritus-feeding isopods (table 8.3). Only 17 percent of the species have high preservation potential. Two poorly represented endobenthic megaguilds (detritus-feeders and carnivores) are entirely lost, being constituted entirely by polychaetes with organic skeletons. Two megaguilds had mineralized species. Endobenthic suspension-feeders were represented by fewer species in the living fauna and as potential fossils than the endobenthic deposit-feeders. In contrast, both the number of individuals and the mass in the living fauna and also as potential fossils were higher for the endobenthic suspension-feeders. Therefore, although three of five megaguilds would be lost during fossilization, the relative abundance of the two dominant megaguilds could potentially be well retained, with only relatively minor change in percentage representation.

LIM CHANNEL

Although Lim Channel (arrowed locality no. 1 in fig. 7.9) is located in a region where the open-water benthos is typically dominated by epibenthos,

TABLE 8.3   Average for Megaguilds and Major Taxonomic Groups Present in Forty-five 0.07 m² Grab Samples at About 15 m–Depth Offshore of Ravenna

Megaguild	Taxonomic Group	No. Species		No. Individuals		Wet Biomass per m²	Estimated Mass of Skeletal Residue
		Live	Mineral	Live	Mineral		
Epibenthic detritus-feeder	Crustacea	1	0	75	0	0.10 g	0 g
Endobenthic suspension-feeder	Bivalvia	1	1	1,001	1,001	27.03	6.22
Endobenthic suspension-feeder	Polychaeta	3	0	130	0	11.55	0
Endobenthic detritus-feeder	Polychaeta	2	0	268	0	0.15	0
Endobenthic deposit-feeder	Bivalvia	2	2	41	41	4.62	1.06
Endobenthic deposit-feeder	Polychaeta	5	0	318	0	5.92	0
Endobenthic carnivore	Polychaeta	4	0	144	0	9.31	0

Based on species and wet biomass data in Crema et al. 1991. Estimated mass of skeletal residue is based on ratios derived from identical or similar species in Zavodnik 1967.

the mud that accumulates on the floor of the channel is inhabited largely by endobenthos. The macrofauna at the locality repeatedly dredged by Zavodnik (1971) was comprised of 51 species, 30 percent of which were well mineralized (table 8.4).

Of the six megaguilds represented, three were epibenthic but had only about one-third of the total species. Each of the epibenthic megaguilds contained a single species with a mineralized skeleton, making 20 percent of the potentially preservable species, but as in the living fauna constituting only a tiny part of the mass of the fossilizable residue. Each endobenthic megaguild was represented by multiple species with mineralized skeletons, but each of them had a different proportion of species with organic skeletons so that the

TABLE 8.4 Megaguilds and Major Taxonomic Groups Present in the Dredged Samples from Zavodnik's Site 601 in Lim Channel

Megaguild	Taxonomic group	No. Species	No. Species in Megaskeletal Residue	Wet Biomass	Mass of Ash Residue
Attached low suspension-feeder	Polychaeta	6	1	0.09+g	0.01 g
Mobile epibenthic detritus-feeder	Gastropoda	1	1	0.04	0.03
Epibenthic carnivore/omnivore	Nemertea	1	0	0.08	0
Epibenthic carnivore/omnivore	Polychaeta	2	0	trace	0
Epibenthic carnivore/omnivore	Crustacea	8	0	0.21	0
Epibenthic carnivore/omnivore	Asteroidea	1	1	0.72	0.12
Endobenthic suspension-feeder	Gastropoda	1	1	1.65	0.79
Endobenthic suspension-feeder	Bivalvia	5	5	4.42	1.38
Endobenthic suspension-feeder	Polychaeta	2	0	trace	0
Endobenthic suspension-feeder	Crustacea	1	0	0.1	0
Endobenthic suspension-feeder	Ophiuroidea	2	2	0.95	0.19
Endobenthic deposit-feeder	Bivalvia	4	0	1.35	0.53
Endobenthic deposit-feeder	Polychaeta	12	1	8.70+	0.64
Endobenthic deposit-feeder	Holothuroidea	2	0	6.49	trace
Endobenthic deposit-feeder	Echinoidea	1	1	9.37	2.86
Endobenthic carnivore	Gastropoda	1	1	13.45	7.58
Endobenthic carnivore	Scaphopoda	1	1	0.23	0.11

relative species richness was reordered in the fossilizable residue. Wet biomass and mineral residue of the endobenthic suspension-feeders comprised almost identical percentages in their respective categories. Endobenthic deposit-feeders had the greatest biomass of all megaguilds in the living fauna and endobenthic carnivores the next-greatest, but the two swapped positions in the mineral residue. Considered as potential fossils, no megaguilds were lost in the Lim Channel fauna, and although some changed their relative positions in species or mass representation, both the living fauna and the potential fossil residues were resolutely endobenthic-dominated.

INTERMEDIATE

GULF OF TRIESTE

Much of the Gulf of Trieste is occupied by a mixed endobenthic and epibenthic fauna (region A in fig. 7.7). Fifty-two species were recognized in a set of 15 grab samples, 33 percent of which were well mineralized (table 8.5). As in Lim Channel, six megaguilds were represented, three of which were epibenthic and had about one-third of the total species. However, in contrast with the fauna sampled in Lim Channel, epibenthos comprised about a third of the readily preservable species, wet biomass, and estimated mass of skeletal residues. The only megaguild not potentially represented by megafossils was mobile epibenthic suspension-feeders. Among the remaining epibenthos, the proportional representation of attached low suspension-feeders increased in the skeletal residues because almost all had mineralized skeletons, and skeletal mass of barnacles comprised much of the wet biomass. Percentage of epibenthic carnivore/omnivore species represented in the skeletal residues is almost identical to that of the living fauna, but high skeletal mass of the gastropods increased the megaguild's representation in the potential skeletal residues. There were relatively small changes from the living fauna to skeletal residues in the proportional representation of species and mass for each of the endobenthic megaguilds. With the exception of loss of mobile suspension-feeders (holothurians), the potential fossil representation of the fauna of the mid-region of the Gulf of Trieste could preserve the ecological structure remarkably well.

In fact, sediments in a transect within the Gulf of Trieste that contains the mixed fauna faithfully reflect at least the epibenthic megaguilds (Zuschin and Perversler 1996). The study focused on the epibenthos, which in the area of the transect—as in the region studied by Fedra et al. (1976)—is concentrated in local "islands" growing on mineralized mollusk and serpulid polychaete skeletons and also on the portions of the organic tubes of the polychaete *Chaetopteris* that project out of the sediment. The islands are host to diverse sedentary organisms (anemones, ascidians, barnacles,

TABLE 8.5 Average for Megaguilds and Major Taxonomic Groups Present in Fifteen 0.2 m² Grab Samples in the Mid-region of the Gulf of Trieste

Megaguild	Taxonomic Group	No. Species	No. Species in Megaskeletal Residue	Wet Biomass per m²	Estimated Mass of Skeletal Residue
Attached low suspension-feeder	Polychaeta	3	2	2.11 g	0.60 g
Attached low suspension-feeder	Gastropoda	1	1	0.01	trace
Attached low suspension-feeder	Crustacea	1	1	1.96	1.77
Mobile suspension-feeder	Holothuroidea	2	0	20.82	trace
Epibenthic carnivore/omnivore	Polychaeta	3	0	0.15	0
Epibenthic carnivore/omnivore	Gastropoda	2	2	24.78	12.14
Epibenthic carnivore/omnivore	Crustacea	4	1	0.50	0.01
Epibenthic carnivore/omnivore	Asteroidea	1	1	0.71	0.18
Endobenthic suspension-feeder	Anthozoa	1	1	4.05	0
Endobenthic suspension-feeder	Bivalvia	3	3	2.90	0.67
Endobenthic suspension-feeder	Gastropoda	1	1	5.84	3.68
Endobenthic suspension-feeder	Polychaeta	7	0	4.19	0
Endobenthic deposit-feeder	Sipunculida	1	0	0.72	0
Endobenthic deposit-feeder	Bivalvia	2	2	0.85	0.20
Endobenthic deposit-feeder	Polychaeta	9	0	9.81	0
Endobenthic deposit-feeder	Holothuroidea	1	0	26.52	0
Endobenthic deposit-feeder	Echinoidea	1	1	48.53	16.01
Endobenthic carnivore	Nemertea	1	0	1.01	0
Endobenthic carnivore	Scaphopoda	1	1	6.07	3.70
Endobenthic carnivore	Polychaeta	7	0	2.94	0

Based on species and wet biomass data in Orel and Mennea 1969. Estimated mass of skeletal residue is based on ratios derived from identical or similar species in Zavodnik 1967.

bryozoans, polychaetes, sponges), as well as mobile organisms (decapods, holothurians, echinoids, ophiuroids). Mineralized organisms attached to chaetopterid tubes tended to disaggregate and accumulate locally upon destruction of the tubes, but mineralized epigrowth on mineralized substrata remained intact after death. Mineralized mobile organisms tended to disaggregate into their individual skeletal elements that were more dispersed within the sediment, but nonetheless were present.

EPIBENTHOS-DOMINATED SITES

FIGAROLA

The macrofauna at this dredged locality (arrowed locality no. 4 on fig. 7.9) was comprised of 73 species, 44 percent of which were well mineralized (table 8.6). Of the seven megaguilds represented, all but two were epibenthic, accounting for 98 percent of the species collected. (Dredging resulted in moderate underrepresentation of endobenthos; see chapter 7.) Both endobenthic megaguilds were poorly represented in the samples and were lost entirely from the skeletal residue (table 8.6). Loss of holothurians from the megaskeletal residue was more serious, because although only 4 percent of the species they accounted for half the biomass, and they were the only epibenthic deposit-feeders present in the material. Other megaguilds were in general more fairly represented in the skeletal residue. Epibenthic attached low suspension-feeders accounted for about half the species and biomass and also about half the species and skeletal mass preservable as megafossils. Epibenthic carnivores/omnivores comprised about one-third of the total species and also of species with skeletons, but their biomass was trivial in the life assemblage. Percent species of three epibenthic megaguilds—carnivores/omnivores, attached erect suspension-feeders, and mobile suspension-feeders—did not change dramatically from the living fauna to the skeletal remains, but each of these groups was overrepresented by an order of magnitude in the skeletal remains, filling the loss of the epibenthic deposit-feeders.

This taxonomic and megaguild reordering of dominance from wet biomass to skeletal remains does not closely reflect the ecological structure of the living community (table 8.6). The living community was dominated by mobile, surficial deposit-feeders and secondarily by attached low suspension-feeders (ascidians), whereas the skeletal remains were actually dominated by primary producers (coralline algae, not included above because they are not animal) and secondarily by attached low suspension-feeders (bryozoans). Neither the holothurians nor the ascidians, which overwhelmingly dominate the mass of the living assemblage, left megascopic skeletal remains, although the holothurians produced large numbers of very small ossicles.

TABLE 8.6   Megaguilds and Major Taxonomic Groups Present in the Dredged Sample from Near Figarola Island

Megaguild	Taxonomic Group	No. Species	No. Species in Megaskeletal Residue	Wet Biomass	Mass of Skeletal Residue
Attached erect suspension-feeder	Porifera	1	0	10.0 g	0 g
Attached erect suspension-feeder	Bryozoa	5	3	7.9	5.1
Attached low suspension-feeder	Porifera	8	0	245.2	0
Attached low suspension-feeder	Cnidaria	4	0	14.3	0
Attached low suspension-feeder	Bryozoa	12	12	110.9	22.1
Attached low suspension-feeder	Polychaeta	3	1	8.4	4.2
Attached low suspension-feeder	Bivalvia	3	3	1.1	0.2
Attached low suspension-feeder	Ascidia	7	0	2,580.5	0
Mobile suspension-feeder	Ophiuroidea	1	1	13.5	3.7
Epibenthic deposit-feeder	Holothuroidea	3	0	3,283.3	0
Epibenthic carnivore/omnivore	Polychaeta	7	0	24.4	0
Epibenthic carnivore/omnivore	Priapulida	1	0	trace	0
Epibenthic carnivore/omnivore	Gastropoda	2	2	6.2	3.5
Epibenthic carnivore/omnivore	Asteroidea	1	1	30.0	4.0
Epibenthic carnivore/omnivore	Arthropoda	13	8	17.5	3.6
Endobenthic suspension-feeder	Polychaeta	1	1	trace	trace
Endobenthic deposit-feeder	Sipunculida	1	0	0.2	0

Banjole

Sections of turf from the *Cellaria* meadow by Banjole Island (fig. 7.15; arrowed locality no. 5 in fig. 7.9) were cut out by hand and contained 54 macrofaunal species, 50 percent of which were potentially preservable as megafossils (table 8.7). The samples collected penetrated into the substratum but contained no endobenthos. Some endobenthos are present where the turf is thin, as evidenced by occasional fecal strings or sediment mounds. However, where the turf is thick as in the samples, the tangle of cuticle-connected proximal segments of *Cellaria* may hinder the penetration of endobenthos both vertically and horizontally below the turf.

TABLE 8.7 Megaguilds and Major Taxonomic Groups Present in a Hand-Collected Sample from Near Banjole Island

Megaguild	Taxonomic Group	No. Species	No. Species in Megaskeletal Residue	Wet Biomass	Mass of Skeletal Residue
Attached erect suspension-feeder	Porifera	1	0	0.1 g	0 g
Attached erect suspension-feeder	Bryozoa	13	10	91.4	49.3
Attached low suspension-feeder	Porifera	7	0	7.0	0
Attached low suspension-feeder	Cnidaria	1	0	trace	0
Attached low suspension-feeder	Bryozoa	12	8	3.3	0.8
Attached low suspension-feeder	Polychaeta	2	1	0.1	0.1
Attached low suspension-feeder	Bivalvia	1	1	trace	trace
Attached low suspension-feeder	Ascidia	4	0	1.8	0
Mobile suspension-feeder	Ophiuroidea	2	2	0.4	0.1
Epibenthic carnivore/omnivore	Polychaeta	4+	0	1.0	0
Epibenthic carnivore/omnivore	Gastropoda	2	2	0.2	0.1
Epibenthic carnivore/omnivore	Arthropoda	5	3	0.8	0.1

Four epibenthic megaguilds were present in the *Cellaria* meadow, none of which were lost from the megaskeletal residue (table 8.7). Half the species richness was included in the attached low suspension-feeding megaguild, a quarter in the attached erect suspension-feeding megaguild, and most of the remaining species were carnivores/omnivores. Percent species representation of the carnivores/omnivores and the mobile suspension-feeders remained virtually unchanged from the living fauna to the skeletal residues, and species of erect and low suspension-feeding were equally well represented in number. Additionally, prolific *Cellaria* and two other robust, well-calcified erect bryozoans caused the erect suspension-feeding megaguild to comprise almost all the biomass and skeletal residue.

The ecological structure of the living community was remarkably well represented in the skeletal remains, which faithfully reflected the disparity in concept of the living community that one would get using only species counts or only biomass. An appreciation of the structure and richness of this community would require knowledge of both the species richness and biomass distribution.

Sediment underlying the *Cellaria* meadow contains skeletal remains that would be predicted from table 8.7. The sediment matrix in cores penetrating the *Cellaria* meadow consists of silt-dominated mud, with sand and coarser grains consisting of carbonate skeletal remains constituting 5–10 percent of the upper 15 cm, overlying a horizon of coralline algal branches in silty mud (McKinney and Jaklin 2001). *Cellaria* branch segments are by far the most numerous megaskeletons (fig. 6.14A, B), followed by other branched bryozoans, encrusting bryozoans, echinoderm ossicles, small gastropods and bivalves, serpulid polychaete tubes, crab claws, and, in the finer sand and silt, abundant sponge spicules.

Sv. Ivan

Sixty species were identified in the macrofaunal sample from this dredged locality (arrowed locality no. 6 of fig. 7.9), 50 percent of which were well mineralized (table 8.8). Of the six megaguilds represented, five were epibenthic and accounted for 95 percent of the species collected. (As for Figarola, dredging resulted in moderate underrepresentation of endobenthos; see chapter 7.) Endobenthic deposit-feeders were poorly represented in the samples and were lost entirely from the skeletal residue (table 8.8). The epibenthic deposit-feeding holothurians were also lost from the megaskeletal residue, which eliminated their megaguild since they were the only representatives, though of low biomass. Three epibenthic megaguilds—attached erect suspension-feeders, attached low suspension-feeders, and epibenthic carnivores/omnivores—each accounted for about 30 percent of the species in the living fauna and also in the skeletal residue. Although rich in species,

TABLE 8.8 Megaguilds and Major Taxonomic Groups Present in the Dredged Sample from Near Sv. Ivan Island

Megaguild	Taxonomic Group	No. Species	No. Species in Megaskeletal Residue	Wet Biomass	Mass of Skeletal Residue
Attached erect suspension-feeder	Porifera	1	0	24.0 g	0 g
Attached erect suspension-feeder	Cnidaria	1	0	1.6	0
Attached erect suspension-feeder	Bryozoa	13	10	2,537.6	853.6
Attached low suspension-feeder	Porifera	7	0	823.7	0
Attached low suspension-feeder	Cnidaria	1	1	2.9	1.7
Attached low suspension-feeder	Bryozoa	7	5	4.6	1.0
Attached low suspension-feeder	Polychaeta	2	2	10.9	7.6
Attached low suspension-feeder	Ascidia	3	0	398.7	0
Mobile suspension-feeder	Ophiuroidea	3	3	8.6	2.3
Epibenthic carnivore/omnivore	Polychaeta	9	0	20.8	0
Epibenthic carnivore/omnivore	Gastropoda	4	4	8.3	4.6
Epibenthic carnivore/omnivore	Asteroidea	1	1	112.0	15.5
Epibenthic carnivore/omnivore	Echinoidea	1	1	3.5	0.1
Epibenthic carnivore/omnivore	Arthropoda	4	3	3.0	0.2
Epibenthic deposit-feeder	Holothuroidea	1	0	219.0	0
Endobenthic deposit-feeder	Echiura	1	0	0.1	0
Endobenthic deposit-feeder	Sipunculida	1	0	0.2	0

the carnivores/omnivores were only a small percent of the biomass and the skeletal residue.

Loss of ascidians and sponges caused the skeletal residue of the attached low suspension-feeders to plummet to 1 percent from the 30 percent biomass that the megaguild had been in the living fauna. The erect suspension-feeding megaguild, comprised largely of robustly calcified bryozoans, rose from about 60 percent of the living fauna to over 95 percent of the total megaskeletal residue. Even with the changes in skeletal mass relative to biomass for the two attached suspension-feeding megaguilds, the skeletal residue of the fauna off Sv. Ivan represents the ecological structure of the living community remarkably well.

Summary

The fossilization potential of the benthos across the northern Adriatic is remarkably good. Even the small samples included in grab samples of 0.2 m^2 generally contain a higher percentage of well-mineralized species than is the norm reported in previous studies. Although grab samples with few specimens have high variation in the proportional representation of number of preservable specimens and of total biomass represented by preservable specimens, larger samples produce ever more consistent results. Except for the polychaetes, which is the most species-rich higher taxon, most of the species-rich, widespread, and abundant taxa are either completely or largely composed of species with mineralized skeletons. Although polychaetes occur virtually ubiquitously and locally in large numbers, most individuals have little biomass compared with other macrobenthos.

In local assemblages for which there are suitable data available, the ecological structure varies from moderately to very well represented by the mineralized species. This is true for both endobenthos-dominated assemblages, in which endobenthic suspension-feeding and deposit-feeding bivalves generally compensate for the loss of polychaetes, and also for epibenthos-dominated assemblages, in which bryozoans and ophiuroids may compensate for the loss of sponges, ascidians, and epibenthic holothurians. There may be complete losses of taxonomic groups within a megaguild in the transition from the living fauna to the potential megaskeletal residues of the fauna. Nonetheless, it is uncommon to lose an entire megaguild from the skeletal residues except instances in which it was minor in both species richness and wet biomass.

There is an interesting and, for this book the most important, consistent result of the conceptual degradation of a series of local faunas from the living communities to their potential fossil residues. In no instance did an endobenthos-dominated fauna change to epibenthos-dominated in the skeletal residues, and in no instance did an epibenthos-dominated fauna change to endobenthos-dominated in the skeletal residues. This encouraging pattern seen for the northern Adriatic benthos gives some confidence that the fossil record represents well the basic ecological structure of ancient benthic faunas, at least to the level of determining whether the fauna was dominated by epibenthos or endobenthos. In addition, the relative abundance of the more important megaguilds probably was commonly preserved more-or-less intact.

CHAPTER NINE

PALEONTOLOGICAL IMPLICATIONS

Die schöpferisch träumende Nature träumte hier und dort dasselbe.

Creatively dreaming, nature dreamed here and there the same dream.
—THOMAS MANN, *Doktor Faustus*

INTRODUCTION

The northern Adriatic was dry land 15,000 years ago and was gradually flooded by the sea between then and a little less than 6,000 years ago. In this geologically brief time the benthic organisms that followed the sea have become organized into diverse local assemblages across the soft-sediment seafloor, with the western reaches rich in endobenthos and the east rich in epibenthos, as seen in chapter 7.

Some regional environmental factor or combination of factors caused this east–west gradient, and this is the topic to be explored in this final chapter, because determination of how the environmental factors relate to the structure of the benthic assemblages here may promote insight into the transition from the Paleozoic to the Modern marine ecosystem.

In this chapter the northern Adriatic benthic fauna is examined in the light of the ecological contrast between the Paleozoic and the Modern as expressed in their prevalent megaguilds. First, Aristocle Vatova's samples are analyzed according to associations of characteristics that define the megaguilds. Then various aspects of the environment and of the samples themselves will be examined to determine how closely the samples correlate with conditions that have been suggested as causes of the change from Paleozoic ecology to Modern ecology. Finally, possible insights into the change from Paleozoic to Modern ecology that can be derived from the northern Adriatic benthos will be summarized.

DESCRIBING LIFE-HABIT ATTRIBUTES

Ultimately this book is about comparison of the benthos of the northern Adriatic Sea with Phanerozoic evolution of the marine ecosystem as described by Richard Bambach (1983). Yet in characterizing benthic species in the northern Adriatic, the terms that I have used only partially overlap with the terms used by Bambach (fig. 1.6). The terms used here and those of Bambach are compared in table 9.1. There are no differences in characterization of benthos as epibenthic versus endobenthic. In characterizing feeding type, I have combined omnivores and herbivores with grazers, and deposit-feeders have been subdivided into surface-feeding detritus-feeders (many of which ingest sediment along with individual food particles) and subsurface deposit-feeders, which are sediment-swallowers feeding largely on comminuted organic matter, bacteria, and perhaps meiofauna.

The greatest differences between the terms used here and those used by Bambach (1983) have to do with mobility and specificity of position relative to the sediment–water interface. For the analyses that follow I have combined some of Bambach's categories. In the case of sedentary epibenthos (low and erect attached, and reclining), I wanted the statistical strength of a larger category and felt that combining the three categories used by Bambach was justified in that all three groups are more characteristic of the Paleozoic fauna than of the Modern fauna. In the case of shallow versus deep endobenthos, I abandoned the attempt to discriminate between the two categories. Even when the concepts "shallow" and "deep" were used in papers on the endobenthic species occurring in the northern Adriatic, there was no consistent depth by which they were differentiated.

Finally, epibenthic mobile organisms are divided here into those that do not disturb the sediment and those that do (bioturbators), and Bambach's term *active* for some endobenthos was changed to "bioturbator." Bioturbators can therefore all be recognized, but their normal living position relative the sediment–water interface and feeding type can still be recognized by using a three-word description, e.g., epibenthic carnivorous bioturbator, or endobenthic deposit-feeding bioturbator.

WHAT ENTITIES ARE BEING COMPARED?

It should be reemphasized that it is the ecological structure of the benthic communities, not the taxa, that is meant here by "Paleozoic and Modern faunas." Brachiopods, stenolaemate bryozoans, and crinoids are present in the Adriatic, but except for very small known areas where stenolaemates or crinoids are abundant, they are not a major part of the wet biomass. Instead, their general ecological equivalents locally flourish: attached and free-lying sedentary bivalves, encrusting and erect cheilostome bryozoans (fig. 7.13), and suspension-feeding ophiuroids that infest the higher portions of attached erect epibenthic organisms or lie directly on the sediment surface (fig. 7.12).

Early Mesozoic ophiuroids were relatively diverse, and dense populations of epibenthic ophiuroids are known from diverse Early Triassic shallow-shelf facies (Twitchell et al. 2005). Aronson in particular has noted that ophiuroids are remnants of the Paleozoic fauna (Aronson 1987, 1989; Aronson and Blake 1997; Aronson and Harms 1985; Aronson and Sues 1987). His studies have focused on dense populations of suspension-feeding ophiuroids lying or preserved as if they were lying directly on the sediment surface. In addition, the feeding posture of suspension-feeding ophiuroids that use tall structures such as erect bryozoan colonies is directly comparable with the stalked echinoderms that were widespread and common in shallow seas during the Paleozoic.

Principal Components Analysis of the Fauna

As seen in Chapter 8, presence of at least one important mineralized species within most megaguilds is important if a skeletal residue reflects the basic ecological structure of a living community. Comparison of tables 8.2 through 8.8 suggests that skeletal (potential fossil) residues of densely sampled local benthic communities across the northern Adriatic do represent the basic ecological structure of the original communities because they have at least one important mineralized species in most megaguilds that were present in life.

Vatova's (1935, 1949) systematic survey across the entire region necessarily sacrificed intensity of local sampling, depending on a single 0.2 m² sample per station, in order to cover the entire area. Therefore there is a greater chance of difference between the living organisms sampled and their potential fossil residues than for the localities discussed in chapter 8, simply because of fine-scale (e.g., meter-scale) geographic variation in species densities. Statistical analysis of ecological patterns in Vatova's data is therefore restricted first to an analysis of the living organisms, followed by an analysis of the potential fossil remains, in order to check for any differences in the patterns.

ALL SPECIES

For each Van Veen grab sample taken, Vatova (1935, 1949) identified all species of megascopic benthos, and the wet biomass and number of individuals of each species were logged. These data for the 280 localities in the open northern Adriatic (figs. 7.1, 7.9), supplemented by data on megaguild to which each species best belongs, plus type of skeleton (table 9.1), were entered into a spreadsheet.[1] Megaguild was determined for many organisms by personal observation, for others from specific information in the published literature or information from specialists, and for some by inference from summaries of functional morphology for certain groups (e.g., Fauchald and Jumars 1979; Warner 1977).

Values for life-mode attributes in table 9.1 were coded for each species as wet biomass (0–500 g). For each station, the log transformed wet biomass for each life mode attribute was summed, resulting in a vector of 10 values, each value representing the relative importance of life-mode attributes for that sample. The total for all attributes within each set indicated by Roman numerals I–III in table 9.1 equals total for log-transformed biomass of each species at the station. For example, endobenthos summed plus epibenthos summed equal total biomass of an individual station, all summed values for all feeding types equal total biomass of that station, etc. (See table 9.2 for examples.) Log-transformed wet biomass distribution and association of life-habit attributes were analyzed using principal components analysis. The first three factors accounted for approximately 72 percent of the variance between samples, seen as trends along the rotated principal components axes (PCAs) (table 9.3).

Plots of rotated principal component axes (PCAs) show clear extensions along the respective axes if all species are included (fig. 9.1A, B). On each graph, data points are assigned to one of two groups (coded by symbols) based solely on their relative proximity to each axis. Many of the samples formed a tight cluster with low loadings near zero and on opposite sides of the origins from the trends (upper left quadrants of fig. 9.1A, B), due to low biomass within the samples. These samples with low biomass are not differentiated by group and are indicated as small black dots.

---

[1] Geological Society of America Data Repository item 2006185, Appendix DR1, environmental characteristics and zoobenthos occurrence, available online at www.geosociety .org/pubs/ft2006.htm, or on request from editing@geosociety.org or Documents Secretary, GSA, P.O. Box 9140, Boulder, CO 80301–9140, USA.

TABLE 9.1 Megaguild and Skeletal Characters of Northern Adriatic Megascopic Benthic Animals as Used in This Book, Compared with Terms Used by Bambach (1983) to Characterize Megaguilds

MEGAGUILD AND SKELETAL CHARACTERS	BAMBACH EQUIVALENTS
I. Position Relative to Sediment surface	I. Position Relative to Sediment Surface
A. Epibenthos	A. Epibenthic
B. Endobenthos	B. Endobenthic
II. Feeding Type	II. Feeding Type
A. Omnivore/grazer/herbivore	A. Grazer
B. Carnivore	B. Carnivore
C. Suspension–feeder	C. Suspension–feeder
D. Detritus–feeder	D. (Surface) deposit–feeder
E. Deposit–feeder	E. (Subsurface) deposit–feeder
III. Mobility	III. Mobility
A. Sedentary	A. Low and erect attached and reclining epibenthos; shallow and deep passive endobenthos
B. Mobile non–sediment disturber	B. Epibenthic mobile grazers; some epibenthic mobile carnivores
C. Bioturbator	C. Some epibenthic carnivores; shallow and deep active endobenthos
IV. Type Skeleton	
A. Organic, or lightly calcified	
B. Organic with spicules or microscopic ossicles	
C. Megaskeletal mineralized	

Principal components axis 1 is defined by high loadings of the attributes sedentary, suspension-feeder, and epibenthic; PCA 2 is defined by high loadings of bioturbator, endobenthic, and deposit-feeder; and PCA 3 is defined by high loadings of omnivore, mobile nonbioturbator, and epibenthic. The samples which formed the trends along PCA 1 and PCA 3 were almost exclusively from stations located in the northeastern Adriatic (fig. 9.2), with very few samples from coastal in the west.

The samples that formed the trend along PCA 2 were largely from stations in the northwestern Adriatic (fig. 9.2), although a few were from along the Istrian coast and a few with low values were intermixed with

TABLE 9.2 Simplified Examples of Life–Habit Attributes Summed for Hypothetical Stations

	STATION		
	1	1	3
Species			
Tellina distorta	1.3	0	2.7
Schizaster canaliferus	5.2	4.1	0
Ophiothrix quinquemaculata	0	3.7	0
Sycon sp.	0	0.8	0
I			
A. Epibenthos	0	4.5	0
B. Endobenthos	6.5	4.1	2.7
II			
A. Omnivore/grazer/herbivore	0	0	0
B. Carnivore	0	0	0
C. Suspension–feeder	0	4.5	0
D. Detritus–feeder	0	0	0
E. Deposit–feeder	6.5	4.1	2.7
III			
A. Sedentary	0	0.8	0
B. Mobile non–sediment disturber	0	3.7	0
C. Bioturbator	6.5	4.1	2.7

TABLE 9.3 Loadings of Megaguild Characteristics on Principal Components Axes (Varimax Rotated Solution) of Wet Biomass of All Species in Vatova's Northern Adriatic Open–Marine Grab Samples

		PCA 1	PCA 2	PCA 3
Cumulative variance		*41%*	*60%*	*72%*
Factor loadings	Sedentary	**−0.91**	0.06	−0.22
	Suspension–feeder	**−0.91**	−0.27	−0.12
	Epibenthos	**−0.80**	−0.04	**−0.54**
	Bioturbator	−0.17	**−0.94**	0.03
	Endobenthos	−0.20	**−0.93**	0.06
	Deposit–feeder	0.05	**−0.72**	−0.42
	Mobile nonbioturbator	−0.34	−0.10	**−0.85**
	Omnivore	−0.15	−0.00	**−0.79**
	Detritus–feeder	−0.35	−0.21	−0.02
	Carnivore	−0.27	−0.31	−0.11

FIGURE 9.1 Rotated principal components axes for megaguild attributes for wet biomass of all benthic animal species in 280 open-water northern Adriatic samples reported by Vatova (1935, 1949). Low biomass samples are indicated by the cluster of dots immediately above and to the left of the crossing of the zero axes. A) PCA 1 and 2; samples with high loadings for epibenthic, sedentary and suspension-feeder form a trend along PCA 1 and are indicated by open circles, and samples with high loadings for endobenthic, bioturbator, and deposit-feeder form a trend along PCA 2 and are indicated by solid circles. B) PCA 2 and 3; samples with high loadings for omnivore and mobile nonbioturbator form a trend along PCA 3 and are indicated by open triangles, and PCA 2 is characterized in part A.

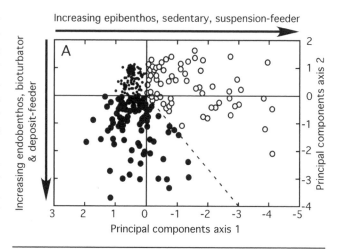

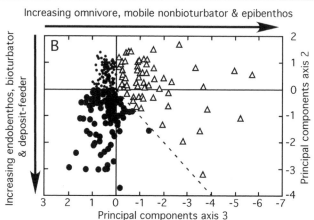

the high PCA 1 stations (fig. 9.2, inset). Note that the relative distance from the origin along a PCA trend is indicated by size of the symbol plotted in figure 9.2. The eastern-most extent of the western high PCA 2 stations can be seen extending to the outer margin of the high PCA 1 belt (fig. 9.2, inset).

Principal components axis 3 identifies a set of samples with high loadings of the attributes epibenthic, mobile nonbioturbating, surface deposit-/detritus-feeding, and omnivorous. These samples are almost exclusively from eastern stations; over half of the high PAC 3 stations correspond with high PCA 1 stations located within the inset in figure 9.2.

Figure 9.2 shows a similar pattern to that of figure 7.3, which was based solely on absolute biomass of endobenthos and epibenthos recorded by Vatova from each station. Grouping of stations by principal components analysis includes, in addition to position relative to the sediment–water face, feeding type, mobility including sediment disturbance, and general skeletal

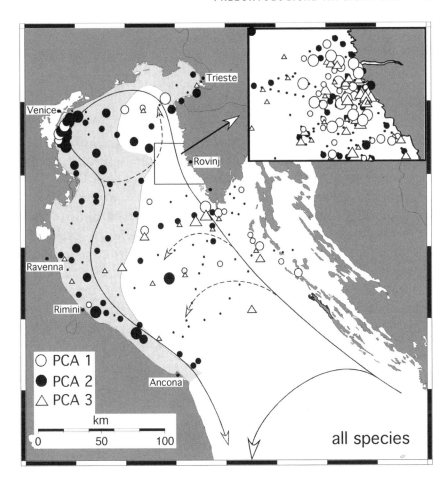

FIGURE 9.2 Distribution of Vatova's (1935, 1949) stations within the northern Adriatic Sea. Stations with samples for which megaguild attributes in fig. 9.1 are affiliated with PCA 1, 2, and 3 are indicated by the same symbols as used in fig. 9.1. Smallest symbols indicate PCA values <1, intermediate-sized symbols indicate intermediate values, and largest symbols indicate PCA values ≥2. The area within the northern Adriatic characterized by average summer chlorophyll *a* values ≥6 μg l⁻¹ from 1948 to 1991 (Zavatarelli et al. 1998) is indicated by light gray shading, whereas the area that averaged less than 6 μg l⁻¹ is not shaded.

characteristics. The loading of attributes on the principal components axes clearly indicates that the western stretches of the northern Adriatic are characterized by a Modern ecology of endobenthic organisms that plow through the sediment, whereas most stations in the northeastern Adriatic are characterized by a Paleozoic ecology of epibenthic sedentary suspension-feeding organisms along with other epibenthic nonbioturbating life habits (McKinney and Hageman 2006).

Endobenthos were present in all but one of Vatova's samples from the 280 stations. Therefore there can be only one of the eastern stations characterized by sedentary, suspension-feeding epibenthos that was completely lacking endobenthic species even though their statistical "signature" is epibenthic.

Mineralized Species

Including the animals with microskeletal elements (sponges, holothurians) and mineralized skeletons with high organic content (numerous crustaceans), there were 195 mineralized species in samples from Vatova's 280 stations in the northern Adriatic. Using the wet biomass of only these species for principal components analysis produces results that differ in detail but that give generally similar results to the analysis for all species. For mineralized species only, the first three factors accounted for approximately 71 percent of the variance between samples, seen as trends along the rotated principal components axes (PCAs) (table 9.4).

Just as for analysis of all species, plots of rotated principal component axes (PCA) show clear extensions along the respective axes when only mineralized species are included (fig. 9.3A, B). Also, many of the samples formed a tight cluster with low loadings near zero and on opposite sides of the origins from the trends (upper left quadrants of fig. 9.1A, B), due to low biomass.

For mineralized species, PCA 1 is defined by high loadings of the attributes mobile nonbioturbator, epibenthos, and omnivore; PCA 2 is defined by high loadings of bioturbator, endobenthos, and deposit-feeder; and PCA

TABLE 9.4 Loadings of Megaguild Characteristics on Principal Components Axes (Varimax Rotated Solution) of Wet Biomass of Mineralized Species in Vatova's Northern Adriatic Open–Marine Grab Samples

		PCA 1	PCA 2	PCA 3
Cumulative variance		*36%*	*59%*	*71%*
Factor loadings	Sedentary	−0.34	0.10	**0.83**
	Suspension–feeder	−0.15	−0.32	**0.87**
	Epibenthos	**−0.77**	−0.04	**0.59**
	Bioturbator	0.04	**−0.96**	0.16
	Endobenthos	0.05	**−0.95**	0.14
	Deposit–feeder	−0.47	**−0.68**	−0.30
	Mobile nonbioturbator	**−0.92**	−0.07	0.14
	Omnivore	**−0.74**	0.02	0.19
	Detritus–feeder	−0.09	−0.27	0.29
	Carnivore	−0.15	−0.34	0.18

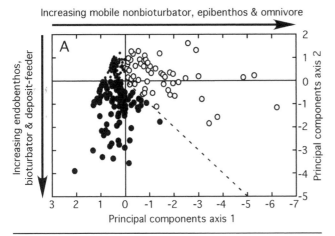

Increasing mobile nonbioturbator, epibenthos & omnivore

FIGURE 9.3 Rotated principal components axes for megaguild attributes for wet biomass of only animal species with mineralized skeletons in 280 open-water northern Adriatic samples reported by Vatova (1935, 1949). Organization of figure is the same as for fig. 9.1.

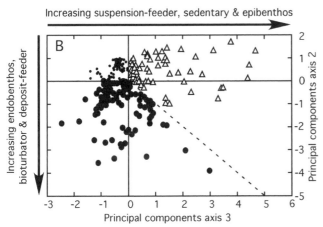

3 is defined by high loadings of suspension-feeder, sedentary, and epibenthos. PCA 2 is unchanged from the analysis for all species (table 9.3), but the loadings for PCA 1 and PCA 3 are reversed. However, both PCA 1 and PCA 3 characterize different aspects of the epibenthos, and the samples that fall along the trends are distributed almost identically as for all species, i.e., almost exclusively from stations located in the northeastern Adriatic (fig. 9.4), with very few samples from coastal stations in the west. The samples which formed the trend along PCA 2 were largely from stations in the northwestern Adriatic (fig. 9.4), although a few were from along the Istrian coast and a few with low values were intermixed with the high PCA 1 stations (fig. 9.4, inset). This pattern, too, is virtually identical to that for the analysis of all species.

There is remarkable overall ecological fidelity of the distributional pattern of potential fossils to living communities in Vatova's data. The loading of attributes for mineralized species on the principal components axes

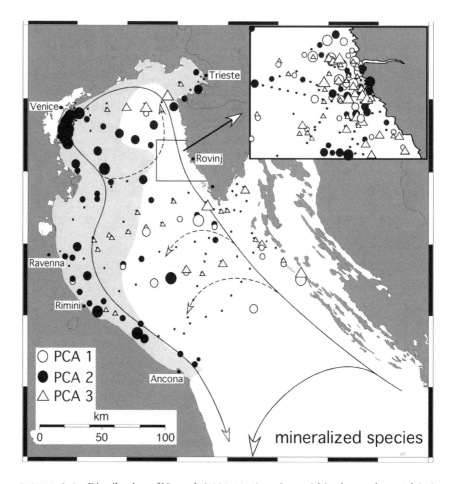

FIGURE 9.4   Distribution of Vatova's (1935, 1949) stations within the northern Adriatic Sea. Stations with samples for which megaguild attributes of mineralized species are affiliated with PCA 1, 2, and 3 in fig. 9.3 are indicated by the same symbols as used in that figure. Symbol size and shading follows the same protocols as in fig. 9.2.

clearly indicates that potential fossils in the western stretches of the northern Adriatic belong among the faunas of the past 200 million years that characterize the Modern marine ecosystem dominated by endobenthic organisms that plow through the sediment. In contrast, the potential fossil residues of most stations in the northeastern Adriatic have ecological affinities with the ancient faunas of the Paleozoic ecosystem dominated by epibenthos.

It is rather surprising that Vatova's samples off the Croatian coast and in some mid-basin stations have such a firmly identifiable Paleozoic ecology, when his reported samples virtually missed entirely the associations of animals that produce such a profoundly Paleozoic visual appearance. The dense communities of ophiuroids (fig. 7.12) and calcified erect bryozoans (figs. 7.13, 7.15) that characterize several regions off the central Istrian coast were apparently sampled but for some reason were not published by Vatova, as described in chapter 7.

CORRELATION OF DISTRIBUTIONAL PATTERNS WITH ENVIRONMENTAL FACTORS

Various long-term environmental changes have been suggested as possible driving forces for the profound taxonomic and ecological shifts from the Paleozoic to the post-Paleozoic. These include orders of magnitude increase in bioturbation, substantial diversification and increase in intensity of predation, and increased nutrient concentration in coastal marine waters. In this section, distribution of endobenthic and epibenthic organisms in the northern Adriatic are examined in the light of these three variables, plus water depth, sedimentation rate, and sediment type.

WATER DEPTH

There is a distinct depth pattern in the distribution of megaguild attributes in the northern Adriatic (table 9.5). At a given depth, however, the megaguild groups commonly differ in biomass significantly from one part of the northern Adriatic to another. In addition to normal depth-related environmental variables such as wave energy, light penetration, and temperature and salinity fluctuations, distribution of epibenthos versus endobenthos apparently is controlled by spatially varying aspects of the environment.

Among the epibenthos, bioturbators are not significantly correlated with depth, but both nonbioturbating mobile benthos and sedentary epibenthos are. Nonbioturbating mobile epibenthos are largely suspension-feeding ophiuroids, e.g., *Ophiothrix quinquemaculata,* and epibenthic suspension- and surface detritus-feeding holothurians, along with diverse non–sediment-probing crustaceans, gastropods, and polychaetes.

Within the sedentary epibenthos, neither megaskeletal nor microskeletal animals are significantly correlated with depth, although the nonmineralized sedentary animals are (Kendal's tau, $p = .002$, $n = 280$). By far the most widely encountered and massive nonmineralized sedentary epibenthos are the ascidians: they tend to occur in relatively deep northern Adriatic waters.

The sedentary and associated mobile epibenthos are relatively most abundant between 30 and 40 m depth (table 9.6), but they are not uniformly distributed across the basin at that depth. Offshore of Istria and

TABLE 9.5 Correlation of Water Depth, Grain Size, and Chlorophyll *a* with Megaguild Attributes in the Northern Adriatic

	Depth		Grain Size		Chlorophyll *a* Summer		Winter	
	tau	*p*	tau	*p*	tau	*p*	tau	*p*
Total biomass	−.179	**<.001**	+.124	.005	.087	.058	.120	.005
Epibenthos	+.061	.139	+.120	**.008**	−.238	**<.001**	−.134	**.002**
Endobenthos	−.287	**<.001**	+.082	.064	+.266	**<.001**	+.253	**<.001**
Sedentary	−.069	.095	+.103	**.023**	−.011	.808	+.015	.738
Mobile								
nonbioturbator	+.069	.101	+.070	.129	−.203	**<.001**	−.111	**.013**
Bioturbator	−.272	**<.001**	+.080	.072	+.252	**<.001**	+.245	**<.001**
Suspension-feeder	−.108	**.008**	+.086	.052	+.095	**.040**	+.135	**.002**
Deposit-feeder	−.138	**.001**	+.037	.404	+.087	.060	+.098	**.023**
Detritus-feeder	−.090	.052	+.116	**.023**	+.127	**.016**	+.112	**.023**
Omnivore, etc.	+.081	.065	+.115	**.017**	−.199	**<.001**	−.164	**<.001**
Carnivore	−.071	.087	+.057	.205	−.034	.471	+.043	.333
Epibenthic *and*								
Sedentary	+.100	**.024**	+.165	**.001**	−.223	**<.001**	−.148	**.002**
Mobile non-								
bioturbator	+.062	.142	+.084	.067	−.192	**<.001**	−.104	**.020**
Suspension-								
feeder	+.126	**.004**	+.163	**.001**	−.258	**<.001**	−.142	**.002**
Omnivore	−.018	.696	+.049	.320	−.177	**.001**	−.128	**.008**
Bioturbator	−.090	.060	−.037	.486	+.092	.092	+.159	**.002**
Endobenthic *and*								
Sedentary	−.177	**<.001**	−.133	**.008**	+.104	**.044**	+.144	**.003**
Bioturbator	−.264	**<.001**	+.090	**.041**	+.25	**<.001**	+.234	**<.001**
Suspension-								
feeder	−.233	**<.001**	−.033	.458	+.315	**<.001**	+.272	**<.001**
Deposit-								
feeder	−.172	**<.001**	+.042	.347	+.119	**.010**	+.138	**.001**

Total benthic biomass and megaguild attributes as correlated with depth, sediment grain size (phi, approximated from sediment descriptions given in Vatova 1935, 1949), and summer and winter chlorophyll *a* concentration in surface water based on averaged data collected 1948–1991 (see fig. 5.6). Biomass data are from 280 stations sampled by Vatova (1935, 1949); chlorophyll a correlations are based on 272 stations, omitting eight stations near or in harbors where nutrient concentrations are unknown.

TABLE 9.6   Ten-Meter Depth Bins of Total Wet Biomass of Epibenthos and of
Endobenthos in Vatova Grab Samples

Depth	Ratio of Percent Wet Biomass to Percent Stations			
	Epibenthos		Endobenthos	
		Mobile		
	Sedentary	Nonbioturbators	Sedentary	Bioturbators
0–10 m (24)	0.003	0.194	0.426	3.153
10.1–20 m (44)	0.176	0.339	0.788	1.142
20.1–30 m (31)	0.627	0.563	0.347	1.319
30.1–40 m (121)	1.887	1.781	1.736	0.655
40.1–50 m (26)	0.311	0.463	0.343	1.053
50.1–60 m (22)	0.685	0.605	0.201	0.250
60.1–70 m (8)	0.153	0.129	0.063	0.068
70.1–80 m (4)	0	0.275	0.109	0.126

Data are given as ratios of percent of total biomass of each category to percent of stations per depth
bin. For example, the 24 stations between 0 and 10 m depth constitute 8.5714 percent of the total of
280 stations, and the 0.98 g total biomass of sedentary epibenthos in the 24 samples constitutes
0.0026 percent of the 3,839.33 g total sedentary epibenthos collected at the 280 stations. The ratio of
0.0026:8.5714 is 0.003:1, given as 0.003 in the table, indicating that sedentary epibenthos in the
samples in depths of 10 m and less is only 0.3 percent of the amount that would have been present
if sedentary epibenthos were uniformly distributed across all depths.

southward along the Croatian coast, sedentary epibenthos averaged 145.65
g m^{-2} between 30 and 40 m depth, whereas in the middle and western parts
of the basin at the same depth the average was 0.40 g m^{-2}.

Depth distributions of both sedentary and bioturbating endobenthos
are significant (table 9.5). The negative correlation of endobenthic bioturbators with depth is due to their disproportionate concentration in depths
of less than 30 m (table 9.6). The area most highly influenced by the Po
plume occurs immediately off the mouth of the Po de Pila and southward
along the Italian coast (figs. 4.2, 6.8). Small bivalves and polychaetes dominate the faunas in these shallow coastal areas, but they occur in such high
abundance that even though individuals have little mass, the population
densities are so high that they add up to high total biomass.

Sedentary endobenthos are primarily suspension-feeding and surface
detritus-accumulating polychaetes, plus anemones (Anthozoa). Both the
polychaetes and anemones live with their primary body mass within the sediment and, when stimulated to do so, can retract their feeding structures below
the sediment–water interface. Sedentary endobenthos do not comprise a large
part of the total biomass and are only marginally more prevalent offshore of
Istria (5 g m^{-2}) than in the middle and western part of the basin (2.5 g m^{-2}).

Both megaskeletal and nonmineralized endobenthos are negatively correlated with depth (Kendal's tau, $p < .001$ for both, $n = 280$); there is no significant correlation for microskeletal endobenthos. The only microskeletal endobenthic animals encountered were burrowing holothurians, which are not a prominent part of the abundant benthos in the vicinity of the Po Delta and plume. The shallow-water endobenthos in the vicinity of the Po Delta are largely bivalves (megaskeletal) and polychaetes (nonmineralized if endobenthic), which account for the two negative correlations.

Although distribution of some of the benthic megaguild attributes is highly correlated with depth, depth per se does not appear to control their distribution. Large east–west geographic variations in average biomass per sample at the preferred depth of most of the attributes that are highly correlated with depth indicate that the fundamental cause of the overall east–west pattern is some other factor(s).

SEDIMENT TYPE

There is little or no support for inferring that sediment texture plays a determining role in the east to west change from epibenthos-rich to endobenthos-rich communities. Epibenthos is abundant and diverse on silt-dominated mud and sand in the northeastern Adriatic and absent to virtually absent on silt-dominated mud to sand substrates of the northwestern Adriatic.

Relatively fine sediments underlie most of the northern Adriatic, largely medium sand to silt-rich mud across the open sea (fig. 6.7). Coarser sediments are largely restricted to the aprons immediately adjacent to rocky headlands or around the bases of steep-sided carbonate islands. Vatova (1935, 1949) recorded sediments coarser than medium sand at only four of his grab sample sites in the open northern Adriatic.

Vatova (1935) thought that his zoocoenoses with abundant epibenthos tracked the sand belt that lies a few kilometers off the Istrian coast (fig. 7.9). Actually, sedentary epibenthos occur on a broad variety of unconsolidated substrates in the northern Adriatic (fig. 9.5A). They are widespread in the eastern regions on sediments ranging from clay-rich muds to medium sand and even occur on the infrequent coarser substrates. Apparently all that is required in the proper environment, regardless of texture of the substrate, is the presence of local hard substrata such as shells or shell fragments on which growth can begin.

Endobenthos also occur abundantly across the range of soft-sediment textures in the northern Adriatic (fig. 9.5B). They too are equally common on sediments ranging from clay-rich muds to medium sand and also can be found in the infrequent coarser substrates.

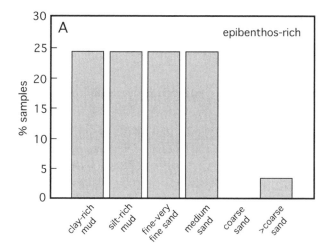

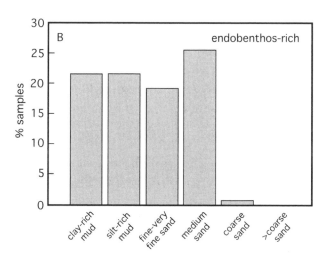

FIGURE 9.5 Basic soft-sediment substrate types for the (A) epibenthos-rich and (B) endobenthos-rich sites indicated in fig. 9.2.

Data on sediment types are from Vatova 1935, 1949.

Grain size of sediments is commonly expressed as a phi scale, which is essentially a method of transforming the power curve represented by the Wentworth size classification of sediments to an arithmetic scoring system. A diameter of 1 mm is assigned a phi value of 0, and in the scale progressively larger diameters are assigned progressively larger *negative* numbers while progressively smaller diameters are assigned progressively larger *positive* numbers. Therefore, if a phenomenon correlates positively with phi, it correlates with decreasing grain sizes. The qualitative terms assigned by Vatova (1935, 1949) to sediment at each station that he occupied were transformed to estimated phi values and entered into the data file for biomass from the stations.

Several attributes of the epibenthos correlate significantly with grain size (table 9.5): epibenthos as a whole, nonbioturbating mobile epibenthos, and sedentary epibenthos. The positive correlations indicate that epibenthos is more prolific on relatively finer-grained than coarser-grained sediments in the northern Adriatic. The inference from Vatova's conclusion that the epibenthos-rich zoocoenoses follow the sand belt offshore of Istria should be tempered (table 7.2); epibenthos in general track finer-grained soft sediments in the northern Adriatic.

Biomass of total endobenthos is not significantly correlated with grain size (table 9.5). However, sedentary endobenthos is significantly correlated negatively with grain size, indicating that they preferentially occur in relatively coarser sediments, perhaps because sedentary tubiculous polychaetes commonly reinforce the walls of their tubes with sand. In contrast, biomass of endobenthic bioturbators correlates positively with phi (table 9.5), indicating that they are more voluminous in finer sediment sizes. However, neither the megaskeletal nor nonmineralized endobenthos correlate at all (Kendal's tau, $p = .952$ and .806, respectively; $n = 280$) whereas microskeletal endobenthos has strong positive correlation (Kendal's tau, $p = .018$, $n = 280$). The only endobenthic microskeletal group is the holothurians, which clearly have a preference for finer-grained sediments. Examination of the distribution of individual taxa or taxonomic groups has demonstrated (e.g., Orel et al. 1987; Schinner 1993) that sediment texture does influence distribution within certain groups, and numerous additional instances of sediment-organism and sediment-community (e.g., Van Hoey et al. 2004) relationships will undoubtedly turn up.

In summary, although distribution of a few of the benthic megaguild attributes correlates moderately well with grain size of the sediment, sediment per se does not appear to control the basinwide distribution pattern of epibenthos versus endobenthos. Large east–west geographic differences in average biomass of megaguilds per sample is due to some other factor(s).

SEDIMENT ACCUMULATION RATE

The highest rates of sediment accumulation in the northern Adriatic probably have an effect on presence, diversity, and abundance of sedentary epibenthos. However, the distribution of faunas with low to absent epibenthic sedentary suspension-feeders is not consistent with this as a primary cause. Sedimentation rates are virtually nil where relict sands occur between Venice and the Gulf of Trieste and mid-basin down to the middle Adriatic, but these are regions with only local pockets rich in sedentary epibenthos (fig. 9.2).

High sedimentation rates are found in the northern Adriatic only near the Po Delta and southward within a few kilometers of the Italian coast (chapter 6). The current 2–4 cm yr^{-1} (Boldrin et al. 1988) or 6.6 g cm^2 yr^{-1} (Frignani et al. 2005) maximum sediment accumulation rate, which is reached only locally near the mouth and along the south side of the Po Delta, is an order of magnitude less than the bryozoan-precluding maximum rate off the Rhône Delta (Lagaaij and Gautier 1965). Sediment accumulation at lower rates does have an effect on types of sedentary suspension-feeding epibenthos within the northern Adriatic. This can be seen in the progressive decrease in diversity and change from sponges and ascidians to byssally attached bivalves from an offshore lower sediment-accumulation site to onshore local higher sediment-accumulation sites on hard substrata near the lagoon of Venice (Gabriele et al. 1999).

Active present-day sediment accumulation is confined to a very narrow belt along the Italian coast north of the Po Delta though broader south of the delta (fig. 6.8), and only 66 of Vatova's stations were located within this region. These stations were assigned to bins depending upon the 0.15–0.30 g cm^2 yr^{-1} interval (Frignani et al. 2005) in which they were located. Not a single attribute or combination of attributes of the endobenthos (table 9.1) was even close to being correlated at $p \leq 0.05$ (Kendal's tau) with sediment accumulation rate in the region. Either an insufficient number of stations were located within the area to pick up any subtle influence of the relatively low sedimentation rate, or few to no stations were located in areas above an accumulation rate threshold.

PREDATORS

The northern Adriatic contains relatively few predators capable of biting, breaking, or scraping mineralized prey such as have been implicated in the Mesozoic marine revolution. Benthic carnivores in the basin include certain gastropods, cephalopods, crustaceans, polychaetes, bottom-feeding fish, and asteroids. Omnivores and grazers found within the gastropods, crustaceans, and epibenthic echinoids also consume benthic animals.

Many surfaces of dead bivalve shells offshore of Istria are covered by fine-scale scratches generated by radulae of grazing gastropods. The gastropods appear to have grazed the bacterial/fleshy-algal growth on bare shell surfaces, because topographically higher organisms with more irregular surfaces, such as encrusting bryozoans, show only trivial breakage or scratches at such fine scale. The radular scratches commonly can be seen up to the bryozoans' edges, where the colonies overgrew grazed surfaces but themselves lack grazed margins or obvious reparative boundaries within the colonies.

Three of the epibenthic echinoid species have strong Aristotle's lanterns that are effective for grazing encrusters off of shell and rock surfaces. Dentition of one of the species is stirodont (*Arbacia lixula*) and in the other two camarodont (*Echinus acutus, Paracentrotus lividus*) (D. Zavodnik, personal communication). Offshore of Rovinj, their five-rayed scraping bite marks are locally abundant on shells covered by a soft film of algae. Field experiments on *A. lixula* and *P. lividus* have shown that grazing on hard substrata by one or both of these two species reduced percent cover of fleshy erect algae but had no effect on abundance of encrusting coralline algae, bryozoans, or barnacles (Bulleri et al. 1999).

The only balistid fish in the region is *Balistes carolinensis*, but it appears to have been uncommon from the time ecosystems were first monitored and is very rarely caught by either bottom trawls or by purse seines (D. Zavodnik, personal communication). Eighteen species of wrasses (family Labridae) occur in the Adriatic (M. Kovačić, personal communication 1999). At least twelve of them occur in the northern Adriatic, where *Coris julis* is rather common (Gamulin-Brida 1974; M. Kovačić, personal communication 1999). These benthic-feeding wrasses primarily eat amphipods, gastropods, and bivalves (Bell and Harmelin-Vivien 1983; Dulčić 1999).

One of the wrasses, *C. julis*, and two species of the sparid fish *Diplodus* are primary predators of juveniles of the grazing echinoids *Arbacia lixula* and *Paracentrotus lividus* on rock substrata (Sala 1977; Guidetti 2004), although paradoxically density and population structure of the predators appears not to affect density of *P. lividus* (Guidetti et al. 2005). Only one species out of twenty fish for which food source could be quantified from an Adriatic rocky slope assemblage, the blenniid *Parablennius gattorugine*, targeted sedentary invertebrates as food although two or three other species incidentally ingested trivial amounts (Fasola et al. 1997). *P. gattorugine*, a dweller of rocky regions, is unlikely to contribute to the deficit of sedentary epibenthos across the muddy northwestern Adriatic seafloor.

Suspension-feeding ophiuroids, especially *Ophiothrix* and *Ophiura* spp., constitute a high proportion of the biomass in many offshore communities in the northeastern Adriatic (e.g., Czihak 1959; Gamulin-Brida et al. 1968; Fedra et al. 1976; Fedra 1977; Ölscher and Fedra 1977; Zuschin et al. 1999), including Zavodnik's station 617 and also the Sv. Ivan locality (fig. 7.9, arrowed locality nos. 6 and 7), prior to the 1989 anoxic event that virtually eliminated the benthos in the deeper waters around Sv. Ivan.

Dense shallow-water populations of ophiuroids can occur where intensity of predation by starfish, crabs, and grazing fish is weak (e.g., Holme 1984; Aronson 1992, 1994; Starmans et al. 1999; Aronson and Blake 2001). Field experiments that compare survival of exposed ophiuroids in low-predation controls versus sites such as rocky reefs where predatory crabs and teleost fish live demonstrate catastrophically high mortality rates by predation in the exposed sites (Aronson and Harms 1985; Aronson 1987, 1989). Starfish are implicated as effective ophiuroid predators on sedimentary substrata (Holme 1984), although their predation rate was lower than that of crabs and teleost fish in a field experiment (Aronson 1989).

Three species of epibenthic ophiuroids examined for arm regeneration rates in and near the Gulf of Trieste were found to average 0.1 to 0.5 scars per arm, which are rates similar to epibenthic species off the Scandinavian coast (Sköld and Rosenberg 1996). *Ophiothrix quinquemaculata* occurred at three of the northern Adriatic study sites, and regeneration rates were significantly different between sites, with more scars per arm at a nearshore site, within a region where pagurid crabs are known to be particularly abundant (Zuschin and Perversler 1996). Within the Gulf of Trieste, where *O. quinquemaculata* preferentially perches on top of the local clumps built by sponges and ascidians, it avoids the clumps and the nearby sedimentary floor occupied by the crab *Pilumnus hirtellus*, one of its predators (Wurzian 1977). The range of ophiuroid arm regeneration rates in the northern Adriatic places predation intensity there with cool-water areas with lower predation intensity as seen in the northeastern Atlantic (Aronson 1989; Sköld and Rosenberg 1996) rather than with the high-intensity predation rates of the tropics, where ophiuroids are absent or in low abundance in shallow open-water areas (Aronson and Harms 1985).

Predation by drilling gastropods is anomalously low off the barrier islands protecting the lagoon of Venice (Kelley 2006), the only area where it has been intensively examined in the northern Adriatic. In a collection of over 1,200 whole mollusk shells, only 12 percent of the included gastropods were predators. Drilling frequency of bivalve shells was 6 percent, and drilling frequency of gastropods was 19 percent, the latter focused largely on *Bittium reticulatum*, which is a species with small individuals. Frequency of incomplete drill-holes relative to all attempts at drilling (complete plus incomplete drill-holes) is used as an index of how effectively the prey escape their drilling gastropod predators. The more highly evolved are the prey's defenses, the greater the proportion of incomplete drill-holes. Only one incomplete drill-hole was found in the entire collection. The low drilling predation rate and the ineffectiveness of the prey in warding off the predators offshore of the lagoon of Venice is consistent with Cretaceous drilling predation, when drilling gastropods were initially radiating, rather than with present-day rates (Kelley 2006).

General predation on epibenthos is apparently not more intense in the western areas of the northern Adriatic where sedentary epibenthos is absent. As one might expect, epibenthic nonbioturbating carnivores actually were taken at higher rates in the eastern samples, where potential epibenthic food (much of it sedentary) was more abundant. This is reflected in the not-quite-significant positive correlation between epibenthic nonbioturbating carnivores and sedentary epibenthos (Kendall's tau = 0.097, $p = 0.06$, $n = 280$).

Epibenthic bioturbators, which are largely carnivores that probe into the sediment for prey, are significantly correlated with endobenthos (Kendall's tau = 0.117, p = 0.014, n = 280). As one would predict, endobenthic carnivores also are significantly correlated with the remaining endobenthos, which includes their prey (Kendall's tau = 0.182, p = 0.049, n = 280).

In the northern Adriatic, predators of epibenthos track distribution of epibenthic prey, and predators of endobenthos track distribution of endobenthic prey. This is hardly surprising. However, predation pressure overall does not appear to be notably intense. In the absence of balistids, the most effective benthic fish predators are the wrasses, which preferentially feed on mobile rather than sedentary epibenthos. The most effective grazing echinoids preferentially reduce fleshy algae rather than mineralized encrusting organisms.

Low predation pressure may be a background condition that allows sedentary epibenthos to thrive in the eastern portion of the northern Adriatic. However, given the absence of negative correlation between predators and prey within the northern Adriatic, the pattern of distribution of sedentary epibenthos there must be a response to other environmental conditions.

BIOTURBATION

It appears that in the northern Adriatic the average density of bioturbators is not sufficient to suppress sedentary epibenthos. Alternatively, the interaction is sufficiently weak that suppression is too weak to account for the east-to-west decrease in sedentary epibenthos.

Numerous bioturbators occur in the northern Adriatic, most notably endobenthic bivalves, polychaetes, gastropods, holothurians, and irregular echinoids, plus epibenthic sediment-penetrating carnivorous gastropods and epibenthic crabs that wedge themselves into the sediment for camouflage. The various bioturbators are widely distributed and virtually ubiquitous across the northern Adriatic: only five of Vatova's 280 grab samples completely lacked bioturbating endobenthos. Epibenthic bioturbators were more sporadic, occurring in only 32 samples and seldom of appreciable biomass.

Sedentary epibenthos were absent in 153, over half, of Vatova's grab samples. One would predict that epibenthos were preferentially lacking from samples with the greatest biomass of bioturbators, but the mean values of biomass of bioturbators were an indistinguishable 80 ± 160 g m^{-2} and 65 ± 115 g m^{-2} for samples respectively lacking and with sedentary epibenthos. There is no significant correlation—either negative or positive—in biomass of bioturbators and sedentary epibenthos (Kendall's tau, p = 0.456, n = 280). Even in the 24 samples where sedentary epibenthos exceeded 250 g m^{-2}, bioturbators were 55 ± 44 g m^{-2}, only slightly less than the basinwide norm.

Focusing on the megaguilds of bioturbators that should churn through sediment most rapidly (endobenthic deposit-feeders, endobenthic carnivores, and endobenthic-feeding epibenthic carnivores), correlations with epibenthic sedentary suspension-feeders are neither negative (suggesting amensalism) nor significant at $p \leq 0.05$ (Kendal's tau, $n = 280$).

Although bioturbators and sedentary epibenthos reach their peak biomass in different parts of the basin, they must be responding to something other than to one another.

NUTRIENT LEVELS

Distributions of megaguild attributes across the northern Adriatic have remarkably high correlations with chlorophyll a concentrations in the overlying water. This appears to be the key for understanding distribution of epibenthos-rich assemblages in the region.

Nutrient levels across the northern Adriatic fluctuate hugely due to the interaction of runoff from the Po and other rivers, wind-driven currents, and summer pooling of low-density water in the shallow regions. Although there can be extreme short-term fluctuations due to windstorms, there is a long-term general pattern as described in chapter 5 and illustrated in figures 5.1–5.3, 5.5, and 5.6.

The primary nutrients (figs. 5.1–5.3) are not immediately available to benthic suspension- and deposit-feeders. They are utilized instead by phytoplankton and photosynthetic bacterioplankton, the standing crop of which is reflected in chlorophyll a concentration in the water. Measurement of chlorophyll a therefore directly reflects the nutrient material available for benthos. Distribution patterns of the primary nutrients and of chlorophyll a are very similar, and in this chapter the terms chlorophyll a and "nutrients" are used interchangeably. Average summer and winter distributions of chlorophyll a for 1948–1991 were assigned to bins as defined by the surface contours drawn in figure 5.6. The summer and winter maps in figure 5.6 were superimposed on the locality map in Vatova (1949; the 1935 map is entirely within one bin), and the bin values for each station were entered into the data file for biomass for Vatova's stations.

There are extensive correlations of attributes of the benthic megaguilds with both summer and winter distribution of chlorophyll a (table 9.5). Total biomass per sample increases with increased chlorophyll a, as does total endobenthos. Microskeletal endobenthos (holothurians) does not correlate with nutrient concentration (Kendal's tau, $p = .811$ and $.481$, summer and winter respectively; $n = 280$), but all other attributes of endobenthic megaguilds that were compared with summer and with winter chlorophyll a are positively correlated. These correlations are highly significant ($p \le .005$), indicating that the endobenthos clearly thrives better in sediments under nutrient-rich water. In this environment the seafloor receives a rich rain of fecal pellets and other detritus that generate organic-rich sediment for deposit-feeders, and endobenthic suspension-feeders have access to plankton-rich water.

With one positive exception, all attributes of epibenthos correlate negatively with summer and winter chlorophyll a (table 9.5). Epibenthos as a whole, mobile epibenthos as a whole (summer only), nonbioturbating mobile epibenthos, and sedentary epibenthos increase significantly as nutrient level decreases. It is probably more informative to state that these attributes of the epibenthos decrease as nutrient level increases.

Bioturbating mobile epibenthos are the one epibenthic group that has a positive correlation with chlorophyll a among the epibenthos. Correlation of epibenthic bioturbators with chlorophyll a is not significant for summer, but the winter correlation is significant at $p = .05$. It is unlikely that the epibenthic bioturbators are directly tracking higher nutrient levels unless some benefit from the presence of higher concentrations of benthic diatoms near the surface of organic-rich sediments. But because most epibenthic bioturbators are carnivores of endobenthic prey they are the one epibenthic group that correlates with abundance of endobenthos. The correlation with high nutrient levels is via their endobenthic prey, which are intervening consumers of water-borne nutrients.

Microskeletal organisms appear to be curiously indifferent to nutrient content of water as well as to depth and grain size. Only three taxonomic groups of megabenthos are microskeletal: holothurians, which include endobenthic deposit-feeders and epibenthic detritus- and suspension-feeders; sponges, which are epibenthic-suspension feeders; and the much less abundant epibenthic suspension-feeding octocorallian anthozoans. Endobenthic holothurians are positively correlated with phi values of grain size, so tend to be more voluminous in relatively fine-grained sediments, but otherwise there are no significant correlations for any of the microskeletal categories with depth, grain size, or seasonal chlorophyll a. Could their broad environmental tolerance be related to ability to accommodate substantial deformation and alteration of direction in movement (holothurians) or growth (sponges, octocorallians)? If so, then why aren't nonmineralized organisms equally indifferent to these environmental conditions? I know of no general hypotheses that address the enigma.

For distribution of megaguild attributes, nutrient level in the overlying water trumps all other aspects of the benthic environment examined here. All but one of the megaguild attributes of endobenthos are significantly correlated positively with nutrient level, meaning that—almost across the board—endobenthic megaguilds increase in biomass with increased nutrient. Conversely, excepting epibenthic carnivores of endobenthos, all megaguild attributes of epibenthos are significantly correlated negatively with nutrient level, so that epibenthic megaguilds thrive in nutrient-deficient areas and do not thrive in nutrient-rich areas of the northern Adriatic.

Implications for Turnover from Paleozoic to Modern Ecology

The distribution of sedentary suspension-feeding epibenthos-rich and of endobenthos-dominated communities across the northern Adriatic is relatively straightforward: sedentary epibenthos is most characteristic toward the (south)east and is virtually absent toward the (north)west, while endobenthos is more ubiquitous but increases in biomass toward the west (figs. 7.3, 9.2), and this carries through in their potential fossil remains (fig. 9.4). As indicated above, depth, sediment texture, sedimentation rate, and bioturbation-generated amensalism are insufficient to account for the overall distribution.

Predation

In the northern Adriatic, the distribution of predators tracks that of their prey, as one would expect. Both endobenthic and epibenthic predators of endobenthos occur where endobenthic prey are present.

Predators of epibenthos are relatively few, and few of them are as effective as their better-equipped tropical counterparts (see above). The voracious, durophagous balistid fish that may be important in reduction of slowly growing calcified epibenthos on tropical and subtropical shelves (Jackson and McKinney 1990) are almost completely absent. Only three species of echinoids have advanced dentition, and durophagous crabs in general appear more focused on the higher nutrient yields of mollusks. As would be expected and as seen for endobenthos, the predators of epibenthos in the northern Adriatic occur where their epibenthic prey are located. They do not patrol the northwestern Adriatic more vigorously than the northeastern Adriatic.

In the end, the northern Adriatic by itself does not allow any insight into the effectiveness of predation as a primary driving force for the change from the epibenthos-rich Paleozoic fauna to the endobenthos-rich Modern fauna. The intensity of predation on the epibenthos is too low across the entire basin to explain the east–west change in abundance of epibenthos. If the predation intensity in the northern Adriatic were as high as that on tropical to subtropical open marine shelves, one can hypothesize but not test that the epibenthos would be vastly reduced.

BIOTURBATION VERSUS SEDENTARY EPIBENTHOS

The distributional patterns of sedentary epibenthos and bioturbators initially appear to be due to amensalism, where vigorous bioturbation, with its attendant sediment resuspension and physical disruption of the sediment surface, reduces or precludes development of suspension-feeders. This phenomenon was noted by Rhoads and Young (1970) and subsequently demonstrated for some epibenthic and tube-dwelling suspension-feeding taxa (Wilson 1991). Thayer (1979, 1983) developed the hypothesis that the increase in bioturbation from Late Paleozoic into the Mesozoic helped reshape the benthic ecosystem by reducing survival of sedentary suspension-feeding benthos on soft sediments (chapter 1).

An extensive set of data on local faunas through the Jurassic shows clear temporal ecological trends: (1) decrease in the relative abundance of epibenthos; (2) increase in the proportion of epibenthos that is mobile; and (3) relative increase in endobenthic bioturbating taxa (Abershan et al. 2006). Abershan et al. could find no abiotic environmental factors that could have been driving these trends. They concluded that the trends were driven by biological interactions, either increase in endobenthic bioturbation causing the decline and increased mobility of epibenthos, or the contemporaneous diversification of predators driving all three trends. They could not disentangle the two hypotheses because the documented trends were consistent with both.

Despite the almost mutually exclusive distribution of sedentary suspension-feeding epibenthos-rich and endobenthos-dominated communities, the northern Adriatic provides little or no support for the hypothesis that biological bulldozing played a major role in eliminating epibenthic sedentary suspension-feeders. Bioturbating endobenthic animals are present throughout the area of abundant sedentary epibenthos, with only slightly diminished biomass that is not statistically different from areas where epibenthos is essentially lacking.

There is no negative correlation between bioturbating endobenthos and sedentary epibenthos (see above).

NUTRIENTS

The strongest and across-the-board most consistent correlations of distribution of epibenthos and endobenthos in the northern Adriatic are with nutrient concentration in the overlying water, as established above. Epibenthos and endobenthos correlate independently of one another with depth and with sediment grain size, but geographical patterns of depth and sediment size distribution only partially correspond with the distributional pattern of epibenthos-rich versus endobenthos-dominated regions of the seafloor. Only nutrient distribution fits the pattern, and it fits remarkably closely.

Average nutrient content of marine waters and rate of nutrient cycling through marine ecosystems appear to have increased through the Phanerozoic (Bambach 1993, 1999; McKinney 1993; Martin 1996; Allmon and Ross 2001). This temporal trend favors the more energy-intensive organisms that characterize the Modern, endobenthos-rich fauna.

Local faunas with living epibenthic, suspension-feeding descendants of the Paleozoic fauna are known to occur in tropical oligotrophic settings (Jackson et al. 1971; Aronson and Harms 1985; Thayer 1989; Thayer et al. 1992). In these instances the Paleozoic "remnants" are within generally oligotrophic regions that are more broadly characterized by members of the Modern fauna. The Paleozoic remnants are restricted to local cryptic or isolated environments. Such occurrences of dense populations of Paleozoic-dominant clades have been interpreted as an escape from highly escalated biotic interactions, which favor the more energy-intensive Modern fauna.

There is no direct analogy here, because the tropical examples mentioned above involve thriving members of the Paleozoic *fauna*, whereas the northeastern Adriatic has a thriving Paleozoic *ecology*, populated largely by members of the Modern fauna. Only suspension-feeding ophiuroids and the much less abundant stenolaemate bryozoans represent the Paleozoic fauna in the northeastern Adriatic.

However, there are notable convergences. In the tropics, the background condition is of low nutrients, and the Paleozoic aspect develops where vigorous biotic interactions, most notably predation, are relaxed. In the northeastern Adriatic, the background condition is of low predation intensity, and the Paleozoic aspect develops where oligotrophic rather than eutrophic conditions exist.

A much closer parallel, actually almost an exact parallel, with the contrasting types of epibenthic assemblages in the northern Adriatic was previously described for the benthos of McMurdo Sound, Antarctica (Dayton and Oliver 1977). McMurdo Sound also has opposite coasts bathed by oligotrophic and eutrophic water, with a similar partitioning between endobenthic assemblages in the higher-nutrient water and epibenthos-rich assemblages in the lower-nutrient water (Dayton and Oliver 1977). Samples in McMurdo Sound were taken at depths (20–40 m) similar to those across most of the northern Adriatic. The range of

TABLE 9.7   Typical Nutrient Concentration and Predation Intensity
Known for Present-Day Oceanic Regions and Inferred for Shelf-Depth
Paleozoic and Modern Faunas

	NUTRIENTS	PREDATION
Endobenthos-dominated		
Shallow tropical (neritic)	Low	High
Shallow temperate (neritic)	High	Variable
Shallow polar (neritic)	High	Low
Eastern McMurdo Sound (neritic)	High	Low
Northwestern Adriatic Sea (neritic)	High	Low
Modern fauna (neritic)	High	High
Sedentary epibenthos-rich		
Deep sea (abyssal, bathyal)	Low	Low
Western McMurdo Sound (neritic)	Low	Low
Northeastern Adriatic (neritic)	Low	Low
Paleozoic fauna	Low	Low

epibenthic higher taxa in the oligotrophic water of McMurdo Sound includes abundant, diverse epibenthic sedentary suspension-feeders similar
to the epibenthos in the oligotrophic waters of the northern Adriatic. Biomass is on average about seven times more abundant under eutrophic water of McMurdo Sound, just as in the Adriatic total biomass of samples is
positively correlated with winter chlorophyll a ($p = 0.005$; for summer
chlorophyll a, $p = 0.058$).

Dayton and Oliver (1977) focused on comparing the horizontal change
from endobenthos-rich assemblages in eutrophic water to epibenthos-rich
assemblages in oligotrophic water with the similar change in structure of
benthic assemblages from neritic to bathyal and abyssal depths. Perhaps
paleobiologists in general have not utilized Dayton and Oliver's observations sufficiently because they did not directly relate them to paleontological trends, and because marine paleobiologists have a tendency to look
primarily to the tropics for modern analogs.

In this book, my focus is on comparing the horizontal change from endobenthos-rich assemblages in eutrophic water to epibenthos-rich assemblages in oligotrophic water with the similar change in structure of benthic
assemblages *through time* at neritic depths. But I think that there are common threads running through the observations in McMurdo Sound, in the

northern Adriatic, from neritic to abyssal depths in modern oceans, and the differences in shelf-depth faunas back through time into the Paleozoic. All the areas of today's oceans that are endobenthos-dominated are characterized by high nutrient levels, high predation rates, or both (table 9.7). The areas of today's oceans that have abundant sedentary suspension-feeding epibenthos living on soft substrata are characterized by *both* low nutrient levels and low predation rates (table 9.7).

Perhaps then this is the key to the transition from the Paleozoic to Modern benthic ecology: an increase in nutrients that made deposit-feeding more worthwhile hand-in-hand with increased predation intensity making life as a sedentary suspension-feeder much more hazardous. The roughly coeval increase in bioturbation may be relatively coincidental to the decrease in epibenthos.

Summary: Energetics in the Northern Adriatic Benthic Fauna, a Connection with Bambach 1999

The shallow northern Adriatic Sea is situated entirely on continental crust of Adria and is surrounded on three sides by subaerial continental material. It therefore qualifies as an epicontinental or epeiric sea. Because it is small on a global scale and because of having two distinctly different benthic ecologies, it is a nearly ideal three-dimensional laboratory for studying the fourth dimension: time.

If one could see the seafloor while on the ferry from Rovinj, on the Istrian coast of Croatia, to Venice, on the sedimentary coast of Italy, it would be like a maritime ecological time-travel from the Paleozoic to the Recent. In essence, the trip begins in oligotrophic water that has been little-modified during its journey from its nutrient-deficient source in the Ionian portion of the western Mediterranean Sea. The latter part of the Rovinj–Venice trip is through a different water body that has formed by mixing the Ionian water as it moves north beyond Istria with the outflow of the Isonzo to Piave rivers that enter the Adriatic along its northern coast. Even more important, water from the nutrient-rich Po River, which enters the Adriatic not far south of Venice, is also included in the mix.

A boundary is always present between the oligotrophic water of the eastern side of the northern Adriatic and the freshened, eutrophic water on the western side. The boundary is more diffuse during the warm months and tighter during the cold months, and the magnitude of the difference fluctuates somewhat depending on the discharge of the rivers and kinetic variables in the sea. But it is always present, and it essentially defines the boundary between the Paleozoic ecology of the epibenthos-rich faunas of the east and the Modern endobenthos-rich faunas of the west.

Bambach (1999) argued that the invasion of land by tracheophytes raised the nutrient level in marine waters both by the flushing of particulate and dissolved organic matter down stream systems and into the oceans and by accelerated dissolution and downstream flushing of inorganic nutrients released into soils by plant-generated acids. He noted that the Devonian and Cretaceous saw two particularly vigorous pulses of the spread and diversification of plants across the land, the first due to the advent of tracheophyte forests and the second due to the origin and diversification of angiosperms. Bambach (1999:134) noted that "the signal of the influence of organic matter from the land on the marine biosphere is the 'halo' of high productivity ringing the ocean basins in the coastal waters and continental shelves. . . . Prior to the invasion of the land and the diversification of land plants, the nutrient supply from runoff from the land would have been much less than today." Using Jack Sepkoski's huge data set of generic ranges, he expected to see but did not find a corresponding pulse in increasing diversity.

Instead, Bambach found correlative ecological responses. During the Devonian the diversity of relatively low-energy predators declined while the diversity of high-energy predators increased. This corresponded with replacement of clades within phyla that had low resistance to predators and that were poorly designed for stability on unstable soft substrates by clades within the same phyla that were better designed. The subsequent Cretaceous pulse of land-plant biomass expansion was coeval with the Mesozoic marine revolution. Burrowing, dominantly deposit-feeding bivalves diversified apparently because of higher concentrations of organic matter in the shallow marine sediment. But the burrowing, dominantly suspension-feeding bivalves diversified while epibenthic suspension-feeding bivalves did not. At the same time, diversity and efficacy of benthic predators increased severalfold (fig. 1.8). The shift of suspension-feeding bivalves into the sediment was apparently the result of increased predation pressure on epibenthic animals.

The northern Adriatic beautifully fits Bambach's concept of terrestrially driven energetic change in the marine biosphere. He laid out a hypothesis for nutrient-driven changes in the marine biosphere through time; in the northern Adriatic it is at present laid out horizontally. The nutrient-rich waters of the northwestern Adriatic are part of the terrestrially influenced "halo" of high productivity, while the water bathing the northeastern Adriatic seafloor is drawn with minimal terrestrial influence from the open, nutrient-deficient eastern Mediterranean.

Even predation pressure in the northern Adriatic is pre-Cenozoic within the context of Bambach's data—and as determined by Kelley (2006). Using Sepkoski's data set on genera, Bambach found that predators comprise an average of 9 percent of marine fossil genera for Jurassic and Early Cretaceous, 12 percent for the Late Cretaceous, and 21 percent for the Cenozoic (actually a rising trend through the Cenozoic, with 24–25 percent characterizing the Neogene). Predatory benthic and demersal genera in Vatova's samples comprise 16 percent if all genera are counted and also 16 percent if only mineralized genera are counted, well below the Cenozoic average. Very few of the northern Adriatic predators are "high energy." Neogastropods are the most numerous of these, but there are few shell-crushing crabs, and the blennies and wrasses are almost benign predators compared with the balistid fish. Almost a third of the predators are asteroids and scaphopods, the latter being a predator of interstitial meiobenthos.

The west to east decrease in nutrient concentration across the northern Adriatic is analogous with a gradient from onshore into an oligotrophic ocean basin, given the trivial influence that the Balkan coast has on the Ionian water as it moves north in the Adriatic. The corresponding west to east gradient in the benthos across the northern Adriatic is remarkably similar to the onshore to offshore gradient in Paleozoic deposits from the Ordovician through the Permian. There was at first a very narrow onshore mollusk-rich belt in the Ordovician, with the remaining areas of shelf depth occupied by brachiopod-, bryozoan-, crinoid-, and coral-rich communities (Jablonski et al. 1983; Sepkoski and Miller 1985). The mollusk-rich belt gradually expanded across the inner shelf through the rest of the Paleozoic. At any time from Ordovician through Permian there was the same "onshore" to "offshore" gradient that exists across the northern Adriatic, from bivalve- and gastropod-rich communities "onshore" in the west to bryozoan-, ophiuroid-, epibenthic bivalve-rich communities "offshore" in the east.

Increasing onshore nutrient levels were spread progressively farther offshore through time. This likely fuelled the widening belt of mollusk-rich communities through the Paleozoic and allowed them to occupy the entire shelf width after the Paleozoic. Even within the mollusk-dominated post-Paleozoic shelves, the mollusks have at least sometimes been organized into an onshore endobenthos-dominated belt and an offshore zone with abundant free-lying epibenthic bivalves (Jablonski and Bottjer 1983).

A further implication to be drawn from the benthic pattern in the northern Adriatic is that most post-Paleozoic Bryozoa-rich limestones may have resulted from local relaxation of the typical post-Paleozoic condition of high predation and also a retreat from high nutrient levels at shelf depth. A well-described case in point is in the Oligocene–Miocene Murray Supergroup in the Murray Basin of southern Australia, where coeval shallow-water fossil assemblages range from mollusk-dominated to epibenthic suspension-feeder– and coralline algal–dominated (Lukasik et al. 2000). At the time the Murray Basin was a roughly equidimensional marine embayment with a shallow sill across the relatively narrow connection with the Southern Ocean, rather analogous with the Adriatic and its marine connection through the Otranto Strait.

The northern half of the Adriatic seafloor fits the concept of an epeiric ramp as defined by Lukasik et al. (2000) for the Murray Basin, but two important conditions differ: the Murray Basin lacked a major siliciclastic input equivalent to the Po River, and there was no equivalent to the current glacio-eustatic sea level changes to alternately flood and drain the Murray Basin on 100 Kyr cycles. Nevertheless, the onshore to offshore Early Miocene sedimentary facies that accumulated in the Murray Basin during times of inferred dry climate are remarkably similar in general to the types of sediments and biological communities (though generally more bryozoan-rich) that one would encounter along a clockwise journey beginning at the mouth of the Po River and staying a few kilometers offshore until reaching the southern shores of the Istrian Peninsula. Some of the Murray Group nearshore facies are rich in siliciclastic silt and mollusk-rich, including the bivalves *Corbula* and *Nucula* that are locally so abundant in the northwestern Adriatic. Further offshore were progressively more erect sedentary suspension-feeders, especially bryozoans similar to those offshore of mid- and southern Istria, including *Cellaria* in all the facies.

Lukasik et al. (2000) interpreted the onshore–offshore succession of lower Miocene biofacies of the Murray Basin to be due in large part to onshore eutrophic conditions to low mesotrophic offshore where the facies were richest in bryozoans and coralline algae. They inferred an alternative condition that produced offshore limestones richer in burrowing echinoids (perhaps analogous with the biocoenosis in the Gulf of Trieste; see chapter 7) to have been more nutrient-rich due to higher terrestrial runoff.

Paleozoic Bryozoa-rich limestones were largely tropical, whereas only 6 percent of post-Paleozoic Bryozoa-rich limestones formed in the tropics and virtually all others developed between 30° and 60° (Taylor and Allison 1998). Bryozoan limestones almost always are rich in erect growth habits. As was recognized by Taylor and Allison, post-Paleozoic bryozoan limestones are largely absent from the tropics because of the more intense predation pressure there relative to cooler waters. Post-Paleozoic bryozoan limestones are not rare—Taylor and Allison cited 95 deposits—except in relation to mollusk-rich sediments. The very low siliciclastic content in Late Cretaceous and Cenozoic bryozoan limestones suggests that the development of these almost flamboyantly epibenthic suspension-feeding faunas with abundant erect bryozoans was outside or on the outer margin of the terrestrial "halo" of river-borne sediments and nutrients. In other words, the platforms and continental surfaces on which they developed were bathed by less nutrient-rich, open ocean waters, within a regional context of low predation pressure.

Today there are regions other than the northern Adriatic where a soft sediment–covered shallow seafloor is populated by an abundant epibenthic fauna, such as the south shelf of Australia, several areas surrounding New

Zealand, and on the Antarctic shelf. However, the northern Adriatic is uniquely valuable for insights into the evolution of benthic ecology through deep time. Within this microcosm of an epeiric sea, there are pronounced environmental gradients that are well documented and continuously studied, there is a large range of endo- and epibenthic communities that can be related to environmental conditions ranging from all-encompassing to detailed scales, and almost the entire northern Adriatic is accessible for direct observation by SCUBA-diving scientists.

LA TRISTEZZA TRA TRIESTE E TÉRMOLI

An Ecosystem Imperiled and Ultimately Doomed

Anche là, nel turbine, nelle onde di cui una trasmetteva all'altra il movimento che aveva trato lei stessa dall'inerzia, un tentativo di sollevarsi che finiva in uno spostamento orizzontale, egli vedeva l'impassibilità del destino. Non v'era colpa, per quanto ci fosse tanto danno.

Also there, in the turmoil, in the waves which one after the other transmitted the movement that they had taken equally for themselves from inertia, in an attempt to pull themselves up that finished in a horizontal displacement, he saw the impassivity of destiny. There was no culprit despite so much damage done.

—ITALO SVEVO, *Senilità*

The northern Adriatic Sea is doomed. It is the nature of epeiric seas that they are ephemeral. Presuming that the alternating Pleistocene interglacial to glacial cycling continues, glaciers will begin rebuilding within a few tens of thousands of years, drawing sea level down sufficiently to drain the entire northern and most of the middle Adriatic. But after that it should flood again, with an unknown number of glacial cycles to follow.

Two more fundamental reasons for the demise of the entire Adriatic Sea are sediment influx and the possible total subduction of the plate on which it is developed. At present approximately 5.5×10^6 tonnes of sediment per year are flushed into the Adriatic, filling the basin from its northwestern end. It is already filled above present sea level for the length of the Po Plain. If the rate of sediment influx remains constant, it will take only about 7.5 million years to fill the nearly 35,000 km^3 volume of the sea to present-day sea level.

It is unlikely that the Adriatic basin will remain at its present size. Subduction of Adria appears to be slowing down and perhaps grinding to a halt. The focal point of the decreased movement is migrating southward away from the northern approach of the Apennines and Alps and is becoming better characterized as an approach of the Apennines and Dinarides. The distance between the Apennines and Dinarides is gradually diminishing, potentially reducing the material overlying the Adriatic basin to remnants of basin fill thrust up above sea level. The slowing rate of tectonic closure of the Adriatic basin suggests that this is unlikely, but still the diminishing size of the basin will interact with the sediment influx to hasten its eventual transition into dry land.

But for now and in the near future, humans have control over the shallow northern Adriatic's precarious ecosystem. From our first, unrecorded entry into the system as a fish predator and continuing through the twentieth and into the twenty-first century, we have been using and affecting it at increasingly intense rates.

In the near future, worldwide production of greenhouse gases will have a minor effect in altering chemical balances within the sea but likely will have a major effect in raising sea level. The upward change in sea level will per se have little effect on the Adriatic except deepening, increasing the area, and shifting benthic communities. Of course increase in sea level will be devastating to low-lying human constructions such as the cities of Venice, Trieste, Ravenna, Rovinj, and Dubrovnik.

Elevated nutrient inputs and fishing techniques are more immediate challenges for the marine environment. We really don't know what the benthic communities of the northern Adriatic were like before human exploitation of and influence on the sea. Perhaps it wasn't radically different from what was observed during most of the twentieth century, because even though some plant-based communities changed dramatically during the twentieth century, the structure of benthic animal communities of the open sedimentary seafloor apparently did not change appreciably from the 1930s to the late 1980s.

Accelerated rates of nutrient influx after World War II increased the frequency and intensity of algal blooms until near the end of the twentieth century. A decrease in the rate of influx over the past several years suggests that regulations on terrestrial nutrient use and release may have partially ameliorated the problem. Late-twentieth-century spread of benthic hypoxia and anoxia into the eastern region of epibenthic communities has been particularly devastating because of the long time required for the communities to be undisturbed while rebuilding. Direct relationship between anthropogenic eutrophication and algal blooms that result in mucilage events that trigger benthic anoxia is difficult to demonstrate. The trigger for such mucilage events appears to be much more complex than a straightforward increase in nutrient concentration.

Harvesting of food from the sea has become more and more intense as the human population has grown. The pernicious effects of this are felt pretty much through the entire marine ecosystem but are notably devastating to coastal ecosystems worldwide. The northern Adriatic is particularly vulnerable to perturbations and long-term change from fishing because the entire region is so shallow. Most devastating of all is bottom trawling, which is used intensively in the northern Adriatic for harvesting scallops.

Divers report that the linear tracks from individual trawl passes crisscross one another intensely in large areas of the eastern and northernmost northern Adriatic. Unfortunately, these are the areas of slow-growing epibenthic organisms and communities that take years or decades to become reestablished after disruption. A few years into recovery from a devastatingly lethal anoxic event a few years earlier, a brief interval of benthic trawling in part of the Gulf of Trieste set the community back to essentially the same state as immediately following the anoxic event. The destabilizing effects of benthic anoxia and of trawling combine as a potent force. Biologists have seen several slow-growing species become progressively less common over the past few decades, though the relative contributions of anoxia-driven mass mortality and of trawl-driven destruction is not clear. Interestingly, communities recover more rapidly following dredging of the endobenthic-dominated seafloor under the Po plume than in the epibenthos-rich northeastern regions.

Accumulated changes have virtually eliminated some benthic communities. Seagrasses were widespread in the northern Adriatic at the beginning of the twentieth century but by the end of the century were reduced to a few isolated patches. Shallow-water kelp forests on rock substrata were also disappearing during that time, being replaced by algal barrens grazed by the echinoid *Paracentrotus lividus*. It is more difficult to determine the causes of these changes than to determine that a combination of anoxic events and trawling has severely damaged epibenthic communities such as that offshore of Piran, Slovenia. Biology of the previous seagrass and kelp bottoms may have been altered by nutrient loading, arrival of human-introduced exotics, or some other environmental change. Quite a number of benthic communities will be stressed once the rapidly spreading, human-introduced alga *Caulerpa* reaches the northern Adriatic from elsewhere in the Mediterranean.

Even the rocky Adriatic coasts of southern Italy and Croatia have been illegally ravaged in quest of the slow-growing, rock-boring edible bivalve *Lithophaga lithophaga*, which is most abundant in vertical rock surfaces within 5–6 m depth. This species became a fad food during the waning years of the twentieth century. To satisfy demand for *L. lithophaga* large stretches of the coasts became shallow quarries, because the bivalves are broken out of the rock underwater with sledgehammers. The resulting rock surfaces become barrens grazed by the sear urchin *Paracentrotus lividus*, with at minimum decades before the *L. lithophaga* community can be reestablished.

Here's the dilemma. Would I refuse to eat *Lithophaga lithophaga* if in a coastal restaurant it were served as an appetizer along with a glass of the local malvasia or verdeca? Probably not, because I would wonder how it tastes. Besides, I have unreservedly and happily eaten a huge range of nektonic and demersal fish from the Adriatic, as well as plenty of mussels, cockles, clams, calamari, octopus, scallops (*Mea culpa, mea maxima culpa*), and diverse crustaceans. The products of the drainage basin of the Po and other northern Italian rivers have contributed to the quality of my life: prosciutto, salami d'asino, parmigiano, piave vecchio, amarone, prosecco, polento, etc. The list could go on and on, but all these stellar agricultural products have one thing in common: some of the phosphates and nitrates used to produce them were washed into the northern Adriatic.

The mere presence of numerous people concentrated locally on a coast is a problem, especially if the offshore waters have low nutrient concentrations such as along coral reefs or in the northeastern Adriatic. Human feces are rich in organic carbon compounds and phosphates, and urine carries abundant nitrogen in the form of urea. The addition of human wastes, whether treated or too commonly dumped untreated into the water, directly increases the nutrient load and encourages the growth of organisms with higher metabolic rates that displace the original inhabitants. Just like the other hundreds of thousands of visitors to the northern Adriatic coast, I have done my little bit while there to nourish the permanent red tides in some harbors as well as the more widespread lush algal growth in other harbors and the high-density populations of *Paracentrotus lividus* that graze on that algae. Come to think of it, is dredging for biological samples any more benign than trawling for scallops? Clearly, I myself am part of the human-induced environmental stress on the northern Adriatic.

Regulations in Italy have now reduced the freedom with which phosphates and nitrates are released into the streams from agriculture and detergents, and there is some hope that the devastating algal blooms that have plagued the Adriatic Italian coasts will diminish. The regulations were put into place because of pressure from the public, who like to go to the coast for recreation. Effective regulations to diminish the devastating effects of benthic trawling will likely be a harder sell. After all, there is public demand for a continuing supply of diverse seafood, much of which is part of the wild benthic fauna. On a beautifully clear blue day, a brick-red sea or a surface film of bubbly mucilage is something everyone wants to disappear, with a return to the sparkling blue sea that they were hoping to experience. On the other hand, if the sky and the sea are both blue, the unseen, chaotically disrupted benthic community offshore that is scored by trawl tracks has little chance to make an impression or to raise the sympathy of someone sitting on the patio of a coastal restaurant, overlooking the scene and anticipating the eminent arrival of a plate of scallops covered by a modicum of delicately herbed olive oil.

Should we be hopeless about the future of the epibenthic communities of the northeastern and northernmost Adriatic? Not necessarily. It will be a hard sell, requiring some mix of public education to broaden acceptance of restriction on harvesting of wild benthic seafood, and probably the willingness of politicians to push and enforce regulations ahead of public consensus. Perhaps the epibenthic communities can in the end be protected. Granted, they will likely be gone anyway when the sea drains off the area in a few tens of thousands of years, but there will be many generations of scientists and recreational divers before that time who can benefit from seeing a thriving, healthy "Paleozoic" benthic ecosystem. Is it too naïve to argue simply that the world would be the poorer without them?

> Like as the waves make towards the pebbled shore,
> So do our minutes hasten to their end,
> Each changing place with that which goes before,
> In sequent toil all forwards do contend.
> Nativity, once in the main of light,
> Crawls to maturity, wherewith being crowned,
> Crooked eclipses 'gainst his glory fight,
> And time that gave doth now his gift confound.
> Time doth transfix the flourish set on youth,
> And delves the parallels in beauty's brow,
> Feeds on the rarities of nature's truth,
> And nothing stands but for his scythe to mow.
> And yet to times in hope my verse shall stand,
> Praising thy worth, despite his cruel hand.
> —WM SHAKESPEARE, *Sonnet* 60

REFERENCES FOR THE EPILOGUE

Bombace 1992; Caddy et al. 1995; Casale et al. 2004; Conti et al. 2002; Degobbis et al. 1997; Devescovi et al. 2005; de Wit and Bendoricchio 2001; Fanelli et al. 1994; Guidetti et al. 2003; Hall-Spencer et al. 1999; Houde et al. 1999; Jackson et al. 2001; Jickells 1998; Morello et al. 2005; Newell and Ott 1999; Ott 1992; Pettine et al. 1998; Pranovi et al. 2000; Scardi et al. 2000; Schinner et al. 1997.

Abershan, M., W. Kiessling, and F. T. Fürsich. 2006. Testing the role of biological interactions in the evolution of mid-Mesozoic marine benthic ecosystems. *Paleobiology* 32:259–277.

Ahel, M., N. Tepic, and S. Terzić. 2005. Spatial and temporal variability of carbohydrates in the northern Adriatic—a possible link to mucilage events. *Science of the Total Environment* 353:239–150.

Aiello, G., S. Bravi, F. Budillon, G. C. Cristofalo, B. D'Argenio, M. De Lauro, L. Ferraro, E. Marsella, F. Molisso, N. Pelosi, M. Sacchi, and M. A. Tramontano. 1995. Marine geology of the Salento shelf (Apulia, south Italy): Preliminary results of a multidisciplinary study. *Giornale di Geologia* 57:17–40.

Alexander, R. R. 1990. Mechanical strength of shells of selected extant articulate brachiopods: Implications for Paleozoic morphologic trends. *Historical Biology* 3:169–188.

Alexander, R. R., and G. P. Dietl. 2003. The fossil record of shell-breaking predation on marine bivalves and gastropods. In P. H. Kelley, M. Kowalewski, and T. A. Hansen, eds., *Predator–Prey Interactions in the Fossil Record*, pp. 141–176. New York: Kluwer Academic/Plenum.

Aljinović, B., I. Blašković, D. Cvijanović, E. Prelogović, D. Skoko, and N. Brdarević. 1984. Correlation of geophysical, geological, and seismological data in the coastal part of Yugoslavia. *Bollettino di Oceanologia Teorica ed Applicata* 2:77–90.

Allmon, W. D., and R. M. Ross. 2001. Nutrients and evolution in the marine realm. In W. D. Allmon and D. J. Bottjer, eds., *Evolutionary Paleoecology*, pp. 105–148. New York: Columbia University Press.

Ambrogi, R., and D. Bedulli. 1983. Descrizione delle comunità macrobentoniche marine di fronte al delta del Po sulla base di due diversi metodi di prelievo. *Atti del Museo Civico di Storia Naturale di Trieste* 35:145–157.

Ambrogi, R., and A. Occhipinti Ambrogi. 1985. The estimation of secondary production of the marine bivalve *Spisula subtruncata* (da Costa) in the area of the Po River Delta. *Marine Ecology* 6:239–250.

———. 1987. Temporal variations of secondary production in the marine bivalve *Spisula subtruncata* off the Po River Delta (Italy). *Estuarine, Coastal, and Shelf Science* 25:369–379.

Ambrogi, R., D. Bedulli, and G. Zurlini. 1990. Spatial and temporal patterns in structure of macrobenthic assemblages: A three-year study in the northern Adriatic Sea in front of the Po River Delta. *Marine Ecology* 11:25–41.

Ambrogi, R., P. Fontana, and I. Sala. 2001. Long term series (1979–93) of macrobenthos data on the soft bottoms in front of the Po River Delta. *Archo Oceanographica and Limnology* 22:167–174.

Amorosi, A., M. Barbieri, F. Castorina, M. L. Colalongo, G. Pasini, and S. C. Vaiani. 1998a. Sedimentology, micropalaeontology, and strontium-isotope dating of a lower-middle Pleistocene marine succession ("Argille Azzurre") in the Romagna Apennines, northeastern Italy. *Bollettino della Società Geologica Italiana* 17:789–806.

Amorosi, A., L. Caporale, U. Cibin, M. L. Colalongo, G. Pasini, F. Ricci Lucchi, P. Severi, and S. C. Vaiani. 1998b. The Pleistocene littoral deposits (Imola Sands) of the northern Apennines foothills. *Giornale di Geologia* 60:83–115.

Amorosi, A., M. L. Colalongo, and F. Fusco. 1999a. Glacio-eustatic control of continental-shallow marine cyclicity from late Quaternary deposits of the southeastern Po Plain, northern Italy. *Quaternary Research* 52:1–13.

Amorosi, A., M. L. Colalongo, G. Pasini, and D. Preti. 1999b. Sedimentary response to Late Quaternary sea-level changes in the Romagna coastal plain (northern Italy). *Sedimentology* 46:99–121.

Amorosi, A., M. C. Centineo, M. L. Colalongo, G. Pasini, G. Sarti, and S. C. Vaiani. 2003. Facies architecture and latest Pleistocene–Holocene depositional history of the Po Delta (Comacchio area), Italy. *Journal of Geology* 111:39–56.

Amorosi, A., M. C. Centineo, M. L. Colalongo, and F. Fiorini. 2005. Millennial-scale depositional cycles from the Holocene of the Po Plain, Italy. *Marine Geology* 222–223:7–18.

Anderson, H., and J. Jackson. 1987. Active tectonics of the Adriatic region. *Geophysical Journal of the Royal Astronomical Society* 91:937–983.

Antoine, D., and A. Morel. 1995. Algal pigment distribution and primary production in the eastern Mediterranean as derived from coastal zone color scanner observations. *Journal of Geophysical Research* 100:16193–16209.

Aronson, R. B. 1987. Predation on fossil and recent ophiuroids. *Paleobiology* 13:187–192.

———. 1989. Brittlestar beds: Low-predation anachronisms in the British Isles. *Ecology* 70:856–865.

———. 1992. Biology of a scale-independent predator–prey interaction. *Marine Ecology Progress Series* 89:1–12.

———. 1994. Scale-independent biological processes in the marine environment. *Oceanography and Marine Biology: An Annual Review* 32: 435–460.

Aronson, R. B., and D. B. Blake. 1997. Evolutionary paleoecology of dense ophiuroid populations. *Paleontological Society Papers* 3:107–119.

———. 2001. Global climate change and the origin of modern benthic communities in Antarctica. *American Zoologist* 41:27–39.

Aronson, R. B., and C. A. Harms. 1985. Ophiuroids in a Bahamian saltwater lake: The ecology of a Paleozoic-like community. *Ecology* 66:1472–1483.

Aronson, R. B., and H.-D. Sues. 1987. The paleoecological significance of an anachronistic ophiuroid community. In C. W. Kerfoot and A. Sih, eds., *Predation: Direct and Indirect Impacts on Aquatic Communities*, pp. 355–366. Hanover, NH: University Press of New England.

Artegiani, A., D. Bregant, E. Paschini, N. Pinardi, F. Raicich, and A. Russo. 1997a. The Adriatic Sea general circulation, part I: Air–sea interactions and water mass structure. *Journal of Physical Oceanography* 27:1492–1514.

———. 1997b. The Adriatic Sea general circulation, part II: Baroclinic circulation structure. *Journal of Physical Oceanography* 27:1515–1532.

Asioli, A., F. Trincardi, A. Correggiari, L. Langone, L. Vigliotti, S. Van Der Kaars, and J. Lowe. 1996. The late-Quaternary deglaciation in the central Adriatic basin. *Il Quaternario* 9:763–770.

Asioli, A., F. Trincardi, J. J. Lowe, D. Aristegui, L. Langone, and F. Oldfield. 2001. Sub-millennial scale climatic oscillations in the central Adriatic during the Late Glacial: Palaeoceanographic implications. *Quaternary Science Reviews* 20:1201–1221.

Ausich, W. I., and D. J. Bottjer. 1982. Tiering in suspension-feeding communities on soft substrata throughout the Phanerozoic. *Science* 216:173–174.

Babbucci, D., T. Tamburelli, M. Viti, E. Mantovani, D. Albarello, F. D'Onza, N. Cenni, and E. Mugnaioli. 2004. Relative motion of the Adriatic with respect to the confining plates: Seismological and geodetic constraints. *Geophysical Journal International* 159:765–775.

Babcock, L. E. 2003. Trilobites in Paleozoic predator–prey systems, and their role in reorganization of Early Paleozoic ecosystems. In P. H. Kelley, M. Kowalewski, and T. A. Hansen, eds., *Predator–Prey Interactions in the Fossil Record*, pp. 55–92. New York: Kluwer Academic/Plenum.

Bader, B. 2000. Life cycle, growth rate, and carbonate production of *Cellaria sinuosa*. In A. Herrera Cubilla and J. B. C. Jackson, eds., *Proceedings of the 11th*

International Bryozoology Association Conference, pp. 136–144. Balboa: Smithsonian Tropical Research Institute.

Bambach, R. K. 1983. Ecospace utilization and guilds in marine communities through the Phanerozoic. In M. J. S. Tevesz and P. M. McCall, eds., *Biotic Interactions in Recent and Fossil Benthic Communities*, pp. 719–746. New York: Plenum.

———. 1985. Classes and adaptive variety: The ecology of diversification in marine faunas through the Phanerozoic. In J. W. Valentine, ed., *Phanerozoic Diversity Patterns*, pp. 191–253. Princeton: Princeton University Press.

———. 1993. Seafood through time: Changes in biomass, energetics, and productivity in the marine ecosystem. *Paleobiology* 19:372–397.

———. 1999. Energetics in the global marine fauna: A connection between terrestrial diversification and change in the marine biosphere. *Geobios* 32:131–144.

Bargagli, A., A. Carillo, G. Pisacane, P. M. Ruti, M. V. Struglia, and N. Tartaglione. 2002. An integrated forecast system over the Mediterranean Basin: Extreme surge prediction in the northern Adriatic Sea. *Monthly Weather Review* 130:1317–1332.

Bartolini, C., R. Caputo, and M. Pieri. 1996. Pliocene–Quaternary sedimentation in the northern Apennine foredeep and related denudation. *Geological Magazine* 133:255–273.

Battaglia, M., M. H. Murray, E. Serpelloni, and R. Bürgmann. 2004. The Adriatic region: An independent microplate within the Africa–Eurasia collision zone. *Geophysical Research Letters* 31:L09605 (4 p.).

Baumiller, T. K., and G. J. Gahn. 2003. Predation on crinoids. In P. H. Kelley, M. Kowalewski, and T. A. Hansen, eds., *Predator–Prey Interactions in the Fossil Record*, pp. 263–278. New York: Kluwer Academic/Plenum.

Behrensmeyer, A. K., F. T. Fürsich, R. A. Gastaldo, S. M. Kidwell, M. A. Kosnik, M. Kowalewski, R. I. Plotnick, R. R. Rogers, and J. Alroy. 2005. Are the most durable shelly taxa also the most common in the marine fossil record? *Paleobiology* 31:607–623.

Bell, J. D., and M. L. Harmelin-Vivien. 1983. Fish fauna of French Mediterranean *Posidonia oceanica* seagrass meadows, 2: Feeding habits. *Tethys* 11:1–14.

Bengston, S., and Yue Z. 1992. Predatorial borings in Late Precambrian mineralized exoskeletons. *Science* 257:367–369.

Benović, A., S. Fonda Umani, A. Malej, and M. Specchi. 1984. Net-zooplankton biomass of the Adriatic Sea. *Marine Biology* 79:209–218.

Bergamasco, A., and M. Gačić. 1996. Baroclinic response of the Adriatic Sea to an episode of Bora wind. *Journal of Physical Oceanography* 26:1354–1369.

Bergamasco, A., V. Filipetto, A. Tomasin, and S. Carniel. 2003. Northern Adriatic general circulation behaviour induced by heat fluxes variations due to possible climatic changes. *Il Nuovo Cimento* 20C:521–533.

Berry, W. B. N. 1968. *Growth of a Prehistoric Time Scale*. San Francisco: W. H. Freeman.

Bertotti, G., V. Picotti, C. Chilovi, R. Fantoni, S. Merlini, and A. Mosconi. 2001. Neogene to Quaternary sedimentary basins in the south Adriatic (Central Mediterranean): Foredeeps and lithospheric buckling. *Tectonics* 20:771–787.

Bertuzzi, A., J. Faganeli, C. Welker, and A. Brambati. 1997. Benthic fluxes of dissolved inorganic carbon, nutrients, and oxygen in the Gulf of Trieste (northern Adriatic). *Water, Air, and Soil Pollution* 99:305–314.

Bianchi, C.N., and C. Morri. 1996. *Ficopomatus* 'reefs' in the Po River Delta (northern Adriatic): Their constructional dynamics, biology, and influences on the brackish-water biota. *Marine Ecology* 17:51–66.

Bill, M., L. O'Dogherty, J. Guex, P.O. Baumgartner, and H. Masson. 2001. Radiolarite ages in Alpine-Mediterranean ophiolites: Constraints on the oceanic spreading and the Tethys-Atlantic connection. *Geological Society of America Bulletin* 113:129–143.

Blake, D.B., and T.E. Guensburg. 1994. Predation by the Ordovician asteroid *Promopalaeaster* on a pelecypod. *Lethaia* 27:235–239.

Boccaletti, M., F. Calamita, G. Deina, R. Gelati, F. Massari, G. Moratti, and F. Ricci Lucchi. 1990. Migrating foredeep–thrust belt system in the northern Apennines and southern Alps. *Palaeogeography, Palaeoclimatology, Palaeoecology* 77:3–14.

Bochdansky, A.B., and G. Herndl. 1992. Ecology of amorphous aggregations (marine snow) in the northern Adriatic Sea, III: Zooplankton interactions with marine snow. *Marine Ecology Progress* 87:135–146.

Boldrin, A., G. Bortoluzzi, F. Frascari, S. Guerzoni, and S. Rabitti. 1988. Recent deposits and suspended sediments off the Po della Pila (Po River, main mouth), Italy. *Marine Geology* 79:159–170.

Bombace, G. 1992. Fisheries of the Adriatic Sea. In G. Colombo, I. Ferrari, V.U. Ceccherelli, and R. Rossi, eds., *Marine Eutrophication and Population Dynamics*, pp. 327–346. Fredensborg, Denmark: Olsen and Olsen.

Bonini, M., C. Tanini, G. Moratti, L. Piccardi, and F. Sani. 2003. Geological and archaeological evidence of active faulting on the Martana Fault (Umbria-Marche Apennines, Italy) and its geodynamic implications. *Journal of Quaternary Science* 18:695–708.

Bonvicini Pagliai, A.M., A.M. Cognetti Varriale, R. Crema, M. Curini Galletti, and R. Vandini Zunarelli. 1985. Environmental impact of extensive dredging in a coastal marine area. *Marine Pollution Bulletin* 16:483–488.

Borsetti, A.M., L. Capotondi, F. Cati, A. Negri, C. Vergnaud-Grazzini, C. Alberini, P. Colantoni, and P.V. Curzi. 1995. Biostratigraphic events and late Quaternary tectonics in the Dosso Gallignani (central–southern Adriatic Sea). *Giornale di Geologia*, ser. 3, 57:41–58.

Bottjer, D.J., and W.I. Ausich. 1986. Phanerozoic development of tiering in soft substrata suspension-feeding communities. *Paleobiology* 12:400–420.

Brambati, A. 1992. Origin and evolution of the Adriatic Sea. In G. Colombo, I. Ferrari, V.U. Ceccherelli, and R. Rossi, eds., *Marine Eutrophication and Population Dynamics*, pp. 327–346. Fredensborg, Denmark: Olsen and Olsen.

Brambati, A., and G. A. Venzo. 1967. Recent sedimentation in the northern Adriatic Sea between Venice and Trieste. *Studi Trentini di Scienze Naturali*, ser. A, 44:202–274.

Brambati, A., M. Ciabatti, G. P. Fanzatti, F. Marabini, and R. Marocco. 1983. A new sedimentological textural map of the northern and central Adriatic Sea. *Bollettino di Oceanologia Teorica ed Applicata* 1:267–271.

Brana, J., and V. Krajcar. 1995. General circulation of the North Adriatic Sea: Results of long-term measurements. *Estuarine and Coastal Shelf Science* 40:421–434.

Bressan, G., and P. Nichetto. 1994. Some observations on the maërl distribution in the northern Adriatic Sea. *Acta Adriatica* 35:15–20.

Brett, C. E. 2003. Durophagous predation in Paleozoic marine benthic assemblages. In P. H. Kelley, M. Kowalewski, and T. A. Hansen, eds., *Predator–Prey Interactions in the Fossil Record*, pp. 401–432. New York: Kluwer Academic/Plenum.

Buljan, M., and M. Zore-Armanda. 1976. Oceanographical properties of the Adriatic Sea. *Oceanography and Marine Biology Annual Review* 14:11–98.

Bulleri, F., L. Benedetti-Cecchi, and F. Cenelli. 1999. Grazing by the sea urchins *Arbacia lixula* L. and *Paracentrotus lividus* Lam. in the northwest Mediterranean. *Journal of Experimental Marine Biology and Ecology* 241:81–95.

Bullimore, B. 1987. Photographic monitoring of subtidal epibenthic communities, 1986: A report to the Nature Conservancy Council. *Skomer Marine Reserve Subtidal Monitoring Project Report* 6:1–70.

Caddy, J. F., R. Refk, and T. Do-Chi. 1995. Productivity estimates for the Mediterranean: Evidence of accelerating ecological change. *Ocean and Coastal Management* 26:1–18.

Cadée, G. C. 1976. Sediment reworking by *Arenicola marina* on tidal flats in the Dutch Wadden Sea. *Netherlands Journal of Sea Research* 10:440–460.

Carbognin, L., P. Teatini, and L. Tosi. 2004. Eustacy and land subsidence in the Venice Lagoon at the beginning of the new millennium. *Journal of Marine Systems* 51:345–353.

———. 2005. Land subsidence in the Venetian area: Known and recent aspects. *Giornale di Geologia Applicata* 1:5–11.

Carminati, E., and C. Doglioni. 2003. Apennines subduction-related subsidence of Venice (Italy). *Geophysical Research Letters* 30(13):1717(4 p.)

Casale, P., L. Laurent, and G. De Metrio. 2004. Incidental capture of marine turtles by the Italian trawl fishery in the north Adriatic Sea. *Biological Conservation* 119:287–295.

Cataletto, B., M. Cabrini, S. Fonda Umani, L. Milani, and M. C. Pavesi. 1993. Variazioni del contenuto in C della biomassa fito-, microzoo- e mesozooplanctonica nel Golfo di Trieste. *Biologia Marina*, suppl. al Notiziario S.I.B.M. 1:141–144.

Cattaneo, A., and F. Trincardi. 1999. The late Quaternary transgressive record in the Adriatic epicontinental sea: Basin widening and facies partitioning. In

K. M. Bergman and J. W. Snedden, eds., *Isolated Shallow Marine Sand Bodies: Sequence Stratigraphic Analysis and Sedimentologic Interpretation*, pp. 127–146. Tulsa, OK: SEPM.

Cattaneo, A., A. Correggiari, L. Langone, and F. Trincardi. 2003. The late-Holocene Gargano subaqueous delta, Adriatic shelf: Sediment pathways and supply fluctuations. *Marine Geology* 193:61–91.

Cattaneo, A., A. Correggiari, T. Marsset, Y. Thomas, B. Marsset, and F. Trincardi. 2004. Seafloor undulation pattern on the Adriatic shelf and comparison to deep-water sediment waves. *Marine Geology* 213:121–148.

Cavaleri, L., L. Bertotti, and N. Tescaro. 1997. The modelled wind climatology of the Adriatic Sea. *Theoretical and Applied Climatology* 56:231–254.

Cavazzoni Galaverni, S. 1972. Distribuzione costiera delle acque dolci continentali nel mare Adriatico (fino alla trasversale Tremiti-Curzola). *Consiglio Nationale delle Ricerche Laboratorio per lo Studio della Dinamica delle Grandi Masse Rapporto Tecnico* 44:1–18.

Cazenave, A., C. Cabanes, K. Dominh, and S. Mangiarotti. 2001. Recent sea level change in the Mediterranean sea revealed by Topex/Poseidon satellite altimetry. *Geophysical Research Letters* 28:1607–1610.

Celet, P. 1973. The Dinaric and Aegean arcs: The geology of the Adriatic. In A. E. M. Nairn, W. H. Kanes, and F. G. Stehli, eds., *The Ocean Basins and Margins. Volume 4A. The Eastern Mediterranean*, pp. 215–261. New York: Plenum Press.

Cermelj, B., A. Bertuzzi, and J. Faganeli. 1997. Modelling of pore water nutrient distribution and benthic fluxes in shallow coastal waters (Gulf of Trieste, northern Adriatic). *Water, Air, and Soil Pollution* 99:435–444.

Channell, J. E. T., B. D'Argenio, and F. Horváth. 1979. Adria, the African promontory, in Mesozoic Mediterranean palaeogeography. *Earth Science Reviews* 15:213–292.

Chiaudani, G., R. Marchetti, and M. Vighi. 1980. Eutrophication in Emilia-Romagna coastal waters (north Adriatic Sea, Italy): A case study. *Progress in Water Technology* 12:185–192.

Cibin, U., E. Spadafora, G. G. Zuffa, and A. Castellarin. 2001. Continental collision history from arenites of episutural basins in the northern Apennines, Italy. *Geological Society of America Bulletin* 113:4–19.

Cocito, S., and F. Ferdeghini. 1998. Marcatura con colorante ed etichettatura: Due metodi per misurare la crescita in briozoi calcificati. *Atti del 12° Congresso della Associazione Italiana di Oceanologia e Limnologia* 2:351–358.

————. 2000. Morphological variations in *Pentapora fascialis* (Cheilostomatida, Ascophorina). In A. Herrera Cubilla and J. B. C. Jackson, eds., *Proceedings of the 11th International Bryozoology Association Conference*, pp. 176–181. Balboa: Smithsonian Tropical Research Institute.

————. 2001. Carbonate standing stock and carbonate production of the bryozoan *Pentapora fascialis* in the north-western Mediterranean. *Facies* 45:25–30.

Cocito, S., S. Sgorbini, and C. N. Bianchi. 1998a. Aspects of the biology of the bryozoan *Pentapora fascialis* in the northwestern Mediterranean. *Marine Biology* 131:73–82.

Cocito, S., F. Ferdeghini, and S. Sgorbini. 1998b. *Pentapora fascialis* (Pallas) [Cheilostomata: Ascophora] colonization of one sublittoral rocky site after seastorm in the northwestern Mediterranean. *Hydrobiologia* 375–376:59–66.

Cocito, S., M. Novosel, and A. Novosel. 2004. Carbonate bioformations around underwater freshwater springs in the north-eastern Adriatic Sea. *Facies* 50:13–17.

Cognetti, G., C. Lardicci, M. Abbiati, and A. Castelli. 2000. The Adriatic Sea and the Tyrrhenian Sea. In C. R. C. Sheppard, ed., *Seas at the Millennium: An Environmental Evaluation,* 1:267–284. Amsterdam: Pergamon.

Colantoni, P., P. Gallignani, and R. Lenaz. 1979. Late Pleistocene and Holocene evolution of the North Adriatic continental shelf (Italy). *Marine Geology* 33: M41–M50.

Colantoni, P., M. Preti, and B. Villani. 1990. Sistema deposizionale e linea di riva olocenica sommersi in Adriatico al largo di Ravenna. *Giiornale di Geologia,* ser. 3, 52:1–18.

Colantoni, P., G. Babbianelli, F. Mancini, and W. Bertoni. 1997. Coastal defence by breakwaters and sea-level rise: The case of the Italian Northern Adriatic Sea. *Bulletin de l'Institut Océanographique, Monaco* 18:133–150.

Conti, A., A. Stefanon, and G. M. Zuppi. 2002. Gas seeps and rock formation in the northern Adriatic Sea. *Continental Shelf Research* 22:2333–2344.

Conway Morris, S. 1977. Fossil priapulid worms. *Special Papers in Palaeontology* 20:1–95.

Cook, P. L., and P. J. Chimonides. 1978. Observations on living colonies of *Selenaria* (Bryozoa, Cheilostomata), I. *Cahiers de Biologie Marine* 19:147–158.

Cornello, M., and A. Manzoni. 1999. Caratterizzazione stagionale degli insediamenti di organismi macrobentonici su substrati sperimentali nel Bacino Centrale della Laguna di Venezia. *Bollettino del Museo Civico di Storia Naturale di Venezia* 49:135–144.

Correggiari, A., M. E. Field, and F. Trincardi. 1996a. Late Quaternary transgressive large dunes on the sediment-starved Adriatic shelf. In M. De Batist and P. Jacobs, eds., *Geology of Siliciclastic Shelf Seas* (Special Publication 117), pp. 155–169. London: Geological Society .

Correggiari, A., M. Roveri, and F. Trincardi. 1996b. Late Pleistocene and Holocene evolution of the North Adriatic Sea. *Il Quaternario* 9:697–704.

Correggiari, A., F. Trincardi, L. Langone, and M. Roveri. 2001. Styles of failure in late Holocene highstand prodelta wedges on the Adriatic shelf. *Journal of Sedimentary Research* 71:218–236.

Coward, M. P., M. De Donatis, S. Mazzoli, W. Paltrinieri, and F.-C. Wezel. 1999. Frontal part of the northern Apennines fold and thrust belt in the Romagna-Marche area (Italy): Shallow and deep structural styles. *Tectonics* 18:559–574.

Cozzi, S., I. Ivanic, G. Catalano, T. Djakovac, and D. Degobbis. 2004. Dynamics of the oceanographic properties during mucilage appearance in the northern Adriatic Sea: Analysis of the 1997 event in comparison to earlier events. *Journal of Marine Systems* 50:223–241.

Crema, R., A. Castelli, and D. Prevedelli. 1991. Long term eutrophication effects on macrofaunal communities in northern Adriatic Sea. *Marine Pollution Bulletin* 22:503–508.

Crescenti, U. 1971. Sul limite Miocene–Pliocene in Italia. *Geologica Romana* 10:1–22.

Crescenti, U., C. d'Amato, A. Balduzzi, and M. Tonna. 1980. Il Pli-Pleistocene del sottosuolo Abruzzese-Marchigiano tra Ascoli Piceno e Pescara. *Geologica Romana* 19:63–84.

Cushman-Roisin, B., M. Gačić, P.-M.-Poulain, and A. Artegiana. 2001. *Physical Oceanography of the Adriatic Sea*. Dordrecht: Kluwer Academic.

Czihak, G. 1959. Vorkommen und Lebensweise der *Ophiothrix quinquemaculata* in der nordlichen Adria bei Rovinj. *Thalassia Jugoslavica* 1:19–27.

Dayton, P. K., and J. S. Oliver. 1977. Antarctic soft-bottom benthos in oligotrophic and eutrophic environments. *Science* 197:55–58.

de Alteriis, G. 1995. Different foreland basins in Italy: Examples from the central and southern Adriatic Sea. *Tectonophysics* 252:349–373.

de Alteriis, G., and G. Aiello. 1993. Stratigraphy and tectonics offshore of Puglia (Italy, southern Adriatic Sea). *Marine Geology* 113:233–253.

Degobbis, D. 1989. Increased eutrophication of the northern Adriatic Sea, second act. *Marine Pollution Bulletin* 20:452–457.

———. A stoichiometric model of nutrient cycling in the northern Adriatic Sea and its relation to regeneration processes. *Marine Chemistry* 29:235–253.

Degobbis, D., and M. Gilmartin. 1990. Nitrogen, phosphorus, and biogenic silicon budgets for the northern Adriatic Sea. *Oceanologica Acta* 13:31–45.

Degobbis, D., N. Smodlaka, I. Pojed, A. Škrivanić, and R. Precali. 1979. Increased eutrophication of the northern Adriatic Sea. *Marine Pollution Bulletin* 10:298–301.

Degobbis, D., S. Fonda Umani, P. Franco, A. Malej, R. Precali, and N. Smodlaka. 1995. Changes in the northern Adriatic ecosystem and the hypertrophic appearance of gelatinous aggregates. *Science of the Total Environment* 165:43–58.

Degobbis, D., R. Precali, I. Ivančić, N. Smodlaka, and S. Kveder. 1997. The importance and problems of nutrient flux measurements to study eutrophication of the northern Adriatic. *Periodicum Biologorum* 99:161–167.

Degobbis, D., A. Malej, and S. Fonda Umani. 1999. The mucilage phenomenon in the northern Adriatic Sea: A critical review of the present scientific hypotheses. *Annali dell'Istituto superiore di Sanità* 35:373–381.

Degobbis, D., R. Precali, I. Ivančić, N. Smodlaka, D. Fuks, and S. Kveder. 2000. Long-term changes in the northern Adriatic ecosystem related to anthropogenic eutrophication. *International Journal of Environment and Pollution* 13:495–533.

Devescovi, M., B. Ozretić, and L. Iveša. 2005. Impact of date mussel harvesting on the rocky bottom structural complexity along the Istrian coast (northern Adriatic, Croatia). *Journal of Experimental Marine Biology and Ecology* 325:134–145.

Dewey, J. F., W. C. Pitman III, W. B. F. Ryan, and J. Bonnin. 1973. Plate tectonics and the evolution of the Alpine system. *Geological Society of America Bulletin* 84:3137–3180.

Dewey, J. F., M. L. Helman, E. Turco, D. H. W. Hutton, and S. D. Knott. 1989. Kinematics of the western Mediterranean. *Geological Society Special Publication* 45:265–283.

de Wit, M., and G. Bendoricchio. 2001. Nutrient fluxes in the Po basin. *Science of the Total Environment* 273:147–161.

Diaz, R. J., and R. Rosenberg. 1995. Marine benthic hypoxia: A review of its ecological effects and the behavioural responses of benthic macrofauna. *Oceanography and Marine Biology Annual Review* 33:245–303.

Di Bucci, D., and S. Mazzoli. 2002. Active tectonics of the Northern Apennines and Adria geodynamics: New data and a discussion. *Journal of Geodynamics* 34:687–707.

Di Bucci, D., S. Mazzoli, O. Nesci, D. Savelli, M. Tramontana, M. De Donatis, and F. Borraccini. 2003. Active deformation in the frontal part of the northern Apennines: Insights from the lower Metauro River basin area (northern Marche, Italy) and adjacent Adriatic off-shore. *Journal of Geodynamics* 36:213–238.

Di Dio, G., S. Lasagna, D. Preti, and M. Sagne. 1997. Stratigrafia dei depositi quaternari della Provincia di Parma. *Bollettino della Società paleontologica Italiana* 36:179–187.

Dietl, G. P. 2003a. Interaction strength between a predator and dangerous prey: *Sinistrofulgur* predation on *Mercenaria*. *Journal of Experimental Marine Biology and Ecology* 289:287–301.

———. 2003b. Coevolution of a marine gastropod predator and its dangerous bivalve prey. *Biological Journal of the Linnean Society* 80:409–436.

Doglioni, C., F. Mongelli, and P. Pieri. 1994. The Puglia uplift (SE Italy): An anomaly in the foreland of the Apenninic subduction due to buckling of a thick continental lithosphere. *Tectonics* 13:1309–1321.

Dominici, S. 2001. Taphonomy and paleoecology of shallow marine macrofossil assemblages in a collisional setting (Late Pliocene–Early Pleistocene, western Emilia, Italy). *Palaios* 16:336–353.

Driscoll, E. G., and R. A. Swanson. 1973. Diversity and structure of epifaunal communities on mollusc valves, Buzzards Bay, Massachusetts. *Palaeogeography, Palaeoclimatology, Palaeoecology* 14:229–247.

Droser, M. L., and D. J. Bottjer. 1988. Trends in depth and extent of bioturbation in Cambrian carbonate marine environments, western United States. *Geology* 16:233–236.

———. 1989. Ordovician increase in extent and depth of bioturbation: Implications for understanding Early Paleozoic ecospace utilization. *Geology* 17:850–852.

———. 1993. Trends and patterns of Phanerozoic ichnofabrics. *Annual Reviews of Earth and Planetary Sciences* 21:205–225.

Droser, M. L., D. J. Bottjer, and P. M. Sheehan. 1997. Evaluating the ecological architecture of major events in the Phanerozoic history of marine invertebrate life. *Geology* 25:167–170.

Duff, P. McL. D., and A. J. Smith. 1992. *Geology of England and Wales*. London: Geological Society.

Dulčić, J. 1999. The diet of the brown wrasse *Labrus merula* (Labridae) in the eastern Adriatic. *Cybium* 23:381–389.

Dworschak, P. C. 2001. The burrows of *Callianassa tyrrhena* (Petagna 1792) (Decapoda: Thalassinidea). *Marine Ecology* 22:153–166.

Edinger, E. N., J. M. Pandolfi, and R. A. Kelley. 2001. Community structure of Quaternary coral reefs compared with Recent life and death assemblages. *Paleobiology* 27:669–694.

Esu, D., and M. Taviani. 1989. Oligohaline mollusc faunas of the Colombacci Formation (upper Messinian) from an exceptional fossil vertebrate site in the Romagna Apennines: Monticino Quarry (Brisighella, n. Italy). *Bollettino della Società paleontologica Italiana* 28:265–270.

Faganeli, J., and G. Herndl. 1991. Dissolved organic matter in the waters of the Gulf of Trieste (northern Adriatic). *Thalassia Jugoslavica* 23:51–63.

Faganeli, J., A. Avčin, N. Faunko, A. Malej, V. Turk, P. Tušnik, B. Vrišer, and A. Vukovič. 1985. Bottom layer anoxia in the central part of the Gulf of Trieste in the late summer of 1983. *Marine Pollution Bulletin* 16:75–78.

Faganeli, J., J. Pezdić, B. Ogorelec, G. Herndl, and T. Dolenec. 1991. The role of sedimentary biogeochemistry in the formation of hypoxia in shallow coastal waters (Gulf of Trieste, northern Adriatic). *Geological Society Special Publication* 58:107–117.

Faganeli, J., J. Pezdić, B. Oborelec, M. Mišič, and M. Najdek. 1994. The origin of sedimentary organic matter in the Adriatic. *Continental Shelf Research* 14:365–384.

Fairbanks, R. G. 1989. A 17,000 year glacio-eustatic sea level record: Influence of glacial melting rates on the Younger Dryas event and deep-ocean circulation. *Nature* 342:637–642.

Fanelli, G., S. Piraino, G. Belmonte, S. Geraci, and F. Boero. 1994. Human predation along Apulian rocky coasts (SE Italy): Desertification caused by *Lithophaga lithophaga* (Mollusca) fisheries. *Marine Ecology Progress Series* 110:1–8.

Fasola, M., L. Canova, F. Foschi, O. Novelli, and M. Bressan. 1997. Resource use by a Mediterranean rocky slope fish assemblage. *Marine Ecology* 18:51–66.

Fauchald, K., and P. A. Jumars. 1979. The diet of worms: A study of polychaete feeding guilds. *Oceanography and Marine Biology Annual Review* 17:193–284.

Favero, V., and R. S. Barbero. 1981. Evoluzione paleoambientale della Laguna di Venezia nell'area archeologica tra Burano e Canale S. Felice. *Società Veneziana di Scienze Naturali Lavori* 6:119–134.

Fedra, K. 1977. Structural features of a North Adriatic benthic community. In B. F. Keegan, P. O. Ceidigh, and P. J. S. Boaden, eds., *Biology of Benthic Organisms*, pp. 233–246. Oxford: Pergamon.

Fedra, K., E. M. Ölscher, C. Scherübel, M. Stachowitsch, and R. S. Wurzian. 1976. On the ecology of a North Adriatic benthic community: Distribution, standing crop, and composition of the macrobenthos. *Marine Biology* 38:129–145.

Ferretti, M., E. Moretti, D. Savelli, A. Stefanon, M. Tramontana, and F.-C. Wezel. 1986. Late Quaternary alluvial sequences in the north-western Adriatic Sea from UNIBOOM profiles. *Bollettino di Oceanologia Teorica ed Applicata* 4:63–72.

Finetti, I. R., M. Boccaletti, M. Bonini, A. Del Ben, R. Geletti, M. Pipan, and F. Sani. 2001. Crustal section based on CROP seismic data across the North Tyrrhenian–North Apennines–Adriatic Sea. *Tectonophysics* 343:135–163.

Fiorini, F., and S. C. Valani. 2001. Benthic foraminifers and transgressive-regressive cycles in the Late Quaternary subsurface sediments of the Po Plain near Ravenna (northern Italy). *Bollettino della Società paleontologica Italiana* 40:357–403.

Fonda Umani, S. 1985. Hydrology and "red tides" in the Gulf of Trieste (North Adriatic Sea). *Oebalia* 11:141–147.

———. 1996. Pelagic production and biomass in the Adriatic Sea. *Scientia Marina* 60 (suppl. 2):65–77.

Fonda Umani, S., G. Honsell, M. Cabrini, and L. Milani. 1985. A tintinnid "bloom" in the Gulf of Trieste (northern Adriatic Sea). *Oebalia* 11:149–156.

Fonda-Umani, S., E. Ghirardelli, and M. Specchi. 1989. *Gli Episodi di "Mare Sporco" nell'Adriatico dal 1729 ai Giorni Nostri*. Trieste: Regione Autonoma Friuli-Venezia Giulia, Direzione regionale dell'Ambiente.

Fonda Umani, S., P. Franco, E. Ghirardelli, and A. Malej. 1992. Outline of oceanography and the plankton of the Adriatic Sea. In G. Colombo, I. Ferrari, V. U. Ceccherelli, and R. Rossi, eds., *Marine Eutrophication and Population Dynamics*, pp. 347–365. Fredensborg, Denmark: Olsen and Olsen.

Fox, J. M., P. S. Hill, T. G. Milligan, and A. Boldrin. 2004. Flocculation and sedimentation on the Po River Delta. *Marine Geology* 203:95–107.

Franco, P., and A. Michelato. 1992. Northern Adriatic Sea: Oceanography of the basin proper and of the western coastal zone. In R. A. Vollenweider, R. Marchetti, and R. Viviani, eds., *Marine Coastal Eutrophication* (Science of the Total Environment, supplement), pp. 35–62. Amsterdam: Elsevier.

Franić, Z. 2005. Estimation of the Adriatic Sea water turnover time using fallout ^{90}Sr as a radioactive tracer. *Journal of Marine Systems* 57:1–12.

Frignani, M., L. Langone, M. Ravaioli, D. Sorgente, F. Alvisi, and S. Albertazzi. 2005. Fine-sediment mass balance in the western Adriatic continental shelf over a century time scale. *Marine Geology* 222–223:113–133.

Fütterer, D., and J. Paul. 1976. Recent and Pleistocene sediments off the Istrian coast (northern Adriatic, Yugoslavia). *Senckenbergiana Maritima* 8:1–21.

Gabriele, M., A. Bellot, D. Gallotti, and R. Brunetti. 1999. Sublittoral hard substrate communities of the northern Adriatic Sea. *Cahiers de Biologie Marine* 40:65–76.

Gačić, M., G. Civitarese, and L. Ursella. 1999. Spatial and seasonal variability of water and biogeochemical fluxes in the Adriatic Sea. In P. Malanotte-Rizzoli and V. N. Eremeev, eds., *The Eastern Mediterranean as a Laboratory Basin for the Assessment of Contrasting Ecosystems*, pp. 335–357. Dordrecht: Kluwer Academic.

Gambini, R., and M. Tozzi. 1996. Tertiary geodynamic evolution of the southern Adria microplate. *Terra Nova* 8:593–602.

Gambolati, G., G. Giunta, M. Putti, P. Teatini, L. Tomasi, I. Betti, M. Morelli, J. Berlamont, D. De Backer, C. Decouttere, J. Monbaliu, C.S. Yu, I. Brøker, E. D. Christensen, B. Elfrink, A. Dante, and M. Gonella. 1998. Coastal evolution of the upper Adriatic Sea due to sea level rise and natural and anthropic land subsidence. In G. Gambolati, ed., *CENAS: Coastline Evolution of the Upper Adriatic Sea Due to Sea Level Rise and Anthropogenic Land Subsidence*, pp. 1–34. Dordrecht: Kluwer Academic.

Gamulin-Brida, H. 1967. The benthic fauna of the Adriatic Sea. *Oceanography and Marine Biology Annual Review* 5:535–568.

———. 1974. Biocoenoses benthiques de la mer Adriatique. *Acta Adriatica* 15(9):1–102.

Gamulin-Brida, H., A. Požar, and D. Zavodnik. 1968. Contributions aux recherches sur la bionomie benthique des fonds meubles de l'Adriatique du nord (II). *Bioloski Glasnik* 21:157–198.

Gatto, P., and L. Carbognin. 1981. The lagoon of Venice: Natural environmental trend and man-induced modification. *Hydrological Sciences Bulletin* 26:379–391.

Gebrande, H., E. Lüschen, M. Bopp, F. Bleibinhaus, B. Lammerer, O. Oncken, M. Stiller, J. Kummerow, R. Kind, K. Millahn, H. Grassl, F. Neubauer, L. Bertelli, D. Borrini, R. Fantoni, C.Pessina, M. Sella, A. Castellarin, R. Nicolich, A. Mazzotti, and M. Bernabini. 2002. First deep seismic reflection images of the Eastern Alps reveal giant crustal wedges and transcrustal ramps. *Geophysical Research Letters* 29(92):1–4.

Ghisetti, F., and L. Vezzani. 1999. Depth and modes of Pliocene–Pleistocene crustal extension of the Apennines (Italy). *Terra Nova* 11:67–72.

———. 2002a. Normal faulting, extension, and uplift in the outer thrust belt of the central Apennines (Italy): Role of the Caramanico fault. *Basin Research* 14:225–236.

———. 2002b. Normal faulting, transcrustal permeability, and seismogenesis in the Apennines (Italy). *Tectonophysics* 348:155–168.

Giani, M., A. Rinaldi, and D. Degobbis. 2005a. Mucilages in the Adriatic and Tyrrhenian Sea: An introduction. *Science of the Total Environment* 353:3–9.

Giani, M., F. Savelli, D. Berto, V. Zangrando, B. Ćosović, and V. Vojvodic. 2005b. Temporal dynamics of dissolved and particulate organic carbon in the northern Adriatic Sea in relation to the mucilage events. *Science of the Total Environment* 353:126–138.

Gilmartin, M., and N. Revelante. 1983. The phytoplankton of the Adriatic Sea: Standing crop and primary production. *Thalassia Jugoslavica* 19:173–188.

———. 1991. Observations on particulate organic carbon and nitrogen concentrations in the northern Adriatic Sea. *Thalassia Jugoslavica* 23:39–49.

Giordani, P., S. Miserocchi, V. Balboni, A. Malaguti, R. Lorenzelli, G. Honsell, and P. Poniz. 1997. Factors controlling trophic conditions in the north-west Adriatic basin: Seasonal variability. *Marine Chemistry* 58:351–360.

Galadini, R. and P. Messina. 2004. Early-Middle Pleistocene eastward migration of the Abruzzi Apennine (central Italy) extensional domain. *Journal of Geodynamics* 37:57–81.Gotsis-Skretas, O., U. Horstmann, and B. Wiryawan. 2000. Cell size structure of phytoplankton communities in relation to physicochemical parameters and zooplankton in a temperate coastal environment. *Archive of Fishery and Marine Research* 48:265–282.

Gradstein, F. M., J. G. Ogg, and A. Smith, eds. 2004. *A Geologic Time Scale* 2004. Cambridge: Cambridge University Press.

Grenerczy, G., G. Sella, S. Stein, and A. Kenyeres. 2005. Tectonic implications of the GPS velocity field in the northern Adriatic region. *Geophysical Research Letters* 32:L16311 (4 p.)

Guidetti, P. 2004. Consumers of sea urchins, *Paracentrotus lividus* and *Arbacia lixula*, in shallow Mediterranean rocky reefs. *Helgoland Marine Research* 58:110–116.

Guidetti, P., S. Fraschetti, A. Terlizzi, and F. Boero. 2003. Distribution patterns of sea urchins and barrens in shallow Mediterranean rocky reefs impacted by the illegal fishery of the rock-boring mollusc *Lithophaga lithophaga*. *Marine Biology* 143:1135–1142.

Guidetti, P., S. Bussotti, and F. Boero. 2005. Evaluating the effects of protection on fish predators and sea urchins in shallow artificial rocky habitats: A case study in the northern Adriatic Sea. *Marine Environmental Research* 59:333–348.

Hall-Spencer, J. M., C. Froglia, R. J. A. Atkinson, and P. G. Moore. 1999. The impact of Rapido trawling for scallops, *Pecten jacobaeus* (L.), on the benthos of the Gulf of Venice. *ICES Journal of Marine Science* 56:111–124.

Hammond, D. E., P. Giordani, G. Montanari, A. Rinaldi, R. Poletti, G. Rovatti, M. Astorri, and M. Ravaioli. 1984. Benthic flux measurements in NW Adriatic coastal waters. *Memorie della Società geologica Italiana* 27:401–407.

Hammond, D. E., P. Giordani, W. M. Berelson, and R. Poletti. 1999. Diagenesis of carbon and nutrients and benthic exchange in sediments of the northern Adriatic Sea. *Marine Chemistry* 66:53–79.

Harding, L. W., Jr., D. Degobbis, and R. Precali. 1999. Production and fate of phytoplankton: Annual cycles and interannual variability. In T. C. Malone,

A. Malej, L. W. Harding Jr., N. Smodlaka, and R. E. Turner, eds., *Ecosystems at the Land–Sea Margin*, pp. 131–172. Washington: American Geophysical Union.

Harper, E. M. 2003. The Mesozoic marine revolution. In P. H. Kelley, M. Kowalewski, and T. A. Hansen, eds., *Predator–Prey Interactions in the Fossil Record*, pp. 433–455. New York: Kluwer Academic/Plenum.

———. 2005. Evidence of predation damage in Pliocene *Apletosia maxima* (Brachiopoda). *Palaeontology* 48:197–208.

Hass, H. 1948. Beitrag zur Kenntnis der Reteporiden. *Zoologica* 101:1–138.

Hayward, P. J., and F. K. McKinney. 2002. Northern Adriatic Bryozoa from the vicinity of Rovinj, Croatia. *Bulletin of the American Museum of Natural History* 270:1–139.

Herndl, G. J. 1988. Ecology of amorphous aggregations (marine snow) in the northern Adriatic Sea, II: Microbial density and activity in marine snow and its implication to overall pelagic processes. *Marine Ecology Progress Series* 48:265–275.

Herndl, G. J., and P. Peduzzi. 1988. The ecology of amorphous aggregations (marine show) in the northern Adriatic Sea, 1: General considerations. *P.S.Z.N.I. Marine Ecology* 9:79–90.

Hoffmeister, A. P., M. Kowalewski, T. K. Baumiller, and R. K. Bambach. 2004. Drilling predation on Permian brachiopods and bivalves from the Glass Mountains, west Texas. *Acta Palaeontologica Polonica* 49:443–454.

Holme, N. A. 1984. Fluctuations of *Ophiothrix fragilis* in the western English Channel. *Journal of the Marine Biological Association of the United Kingdom* 64:351–378.

Houde, E. D., S. Jukic-Peladic, S. B. Brandt, and S. D. Leach. 1999. Fisheries: Trends in catches, abundance, and management. In T. C. Malone, A. Malej, L. W. Harding Jr., N. Smodlaka, and R. E. Turner, eds., *Ecosystems at the Land–Sea Margin*, pp. 341–366. Washington: American Geophysical Union.

Hrs-Brenko, M. 1981. Population studies on *Corbula gibba* (Olivi), Bivalvia, Corbulidae, in the northern Adriatic Sea. *Journal of Molluscan Studies* 47:17–24.

Hrs-Brenko, M., D. Medaković, Z. Labura, and E. Zahtila. 1994. Bivalve recovery after a mass mortality in the autumn of 1989 in the northern Adriatic Sea. *Periodicum Biologorum* 96:455–458.

Hrvatović, H., and H. Pamić. 2005. Principal thrust-nappe structures of the Dinarides. *Acta Geologica Hungarica* 48:133–151.

Hsü, K. J., W. B. F. Ryan, and M. B. Cita. 1973. Late Miocene desiccation of the Mediterranean. *Nature* 242:240–244.

Ivančić, I., and D. Degobbis. 1987. Mechanisms of production and fate of organic phosphorus in the northern Adriatic Sea. *Marine Biology* 94:117–125.

Jablonski, D. 2000. Micro- and macroevolution: Scale and hierarchy in evolutionary biology and paleobiology. In D. H. Erwin and S. L. Wing, eds., *Deep Time* (Supplement to Paleobiology 26.4), pp. 15–52. Lawrence, KA: Paleontological Society.

Jablonski, D., and D. J. Bottjer. 1983. Soft-bottom epifaunal suspension-feeding assemblages in the Late Cretaceous. In M. J. S. Tevesz and P. L. McCall, eds., *Biotic Interactions in Recent and Fossil Benthic Communities*, pp. 747–812. New York: Plenum.

Jablonski, D., J. J. Sepkoski Jr., D. J. Bottjer, and P. M. Sheehan. 1983. Onshore–offshore patterns in the evolution of Phanerozoic shelf communities. *Science* 222:1123–1125.

Jackson, J. B. C., and F. K. McKinney. 1990. Ecological processes and progressive macroevolution of marine clonal benthos. In R. M. Ross and W. D. Allmon, eds., *Causes of Evolution: A Paleontological Perspective*, pp. 173–209. Chicago: University of Chicago Press.

Jackson, J. B. C., T. F. Goreau, and W. D. Hartman. 1971. Recent brachiopod-coralline sponge communities and their paleoecological significance. *Science* 173:623–625.

Jackson, J. B. C., M. X. Kirby, W. H. Berger, K. A. Bjorndal, L. W. Botsford, B. J. Bourque, R. H. Bradbury, R. Cooke, J. Erlandson, J. A. Estes, T. P. Hughes, S. Kidwell, C. B. Lange, H. S. Lenihan, J. M. Pandolfi, C. H. Peterson, R. S. Steneck, M. J. Tegner, and R. R. Warner. 2001. Historical overfishing and the recent collapse of coastal ecosystems. *Science* 293:629–638.

Jaklin, A. 2002. *Recovery of North Adriatic Sedimentary Bottom Macrofauna after Oxygen Depletion* [in Croatian]. Ph.D. thesis, Zagreb: University of Zagreb Faculty of Science Department of Biology.

Jaklin, A., and E. Zahtila. 1990. 1989 anoxia and mass mortality of macrobenthos in the northern Adriatic. *1ˢᵗ International Symposium "Ecological problems in the Adriatic Sea," 7–9 November 1990 Abstracts*, pp. 144–145. Split.

Jenkins, C., F. Trincardi, L. Hatchett, A. Niedoroda, J. Goff, R. Signell, and K. McKinney. 2005. *Il fondo marino Adriatico*: http://instaar.colorado.edu/~jenkinsc/dbseabed/adr_gis/

Jickells, T. D. 1998. Nutrient biogeochemistry of the coastal zone. *Science* 281:217–222.

Jolivet, L., and C. Faccenna. 2000. Mediterranean extension and the Africa-Eurasia collision. *Tectonics* 19:1095–1106.

Jordan, T. E., and I. Valiela. 1982. A nitrogen budget of the ribbed mussel, Geukensia demissa, and its significance in nitrogen flow in a New England salt marsh. *Limnology and Oceanography* 27:75–90.

Justić, D. 1991. Hypoxic conditions in the northern Adriatic Sea: Historical development and ecological significance. In R. V. Tyson and T. H. Pearson, eds., *Modern and Ancient Continental Shelf Anoxia* (Special Publication 58), pp. 95–105. London: Geological Society.

Kabat, A. R. 1990. Predatory ecology of naticid gastropods with a review of shell boring predation. *Malacologia* 32:155–193.

Kase, T., and M. Ishikawa. 2003. Mystery of naticid predation history solved: Evidence from a "living fossil" species. *Geology* 31:403–406.

Kastens, K., and J. Mascle. 1990. Scientific results of ODP Leg 107. *Proceedings of the Ocean Drilling Project, Scientific Results* 107:3–26.

Kelley, P. H. 2006. Low frequency of drilling predation in the northwest Adriatic: Cretaceous rather than Paleozoic analog. *Geological Society of America Abstracts with Programs* 38(3):22.

Kelley, P. H., and T. H. Hansen. 1993. Evolution of the naticid gastropod predator–prey system: An evaluation of the hypothesis of escalation. *Palaios* 8:358–375.

———. 2003. The fossil record of drilling predation on bivalves and gastropods. In P. H. Kelley, M. Kowalewski, and T. A. Hansen, eds., *Predator–Prey Interactions in the Fossil Record*, pp. 113–139. New York: Kluwer Academic/Plenum.

Kidwell, S. M. 2005. Shell composition has no net impact on large-scale evolutionary patterns in mollusks. *Science* 307:914–917.

Kidwell, S. M., and P. J. Brenchley. 1996. Evolution of the fossil record: Thickness trends in marine skeletal accumulations and their implications. In D. Jablonski, D. H. Erwin, and J. H. Lipps, eds., *Evolutionary Paleobiology*, pp. 290–336. Chicago: University of Chicago Press.

Kollmann, H., and M. Stachowitsch. 2001. Long-term changes in the benthos of the northern Adriatic Sea: A phototransect approach. *Marine Ecology* 22:135–154.

Kourafalou, V. H. 1999. Process studies on the Po River plume, North Adriatic Sea. *Journal of Geophysical Research* 104:29963–29985.

———. 2001. River plume development in semi-enclosed Mediterranean regions: North Adriatic Sea and northwestern Aegean Sea. *Journal of Marine Systems* 30:181–205.

Kovačević, V., M. Gačić, and P.-M. Poulain. 1999. Eulerian current measurements in the Strait of Otranto and in the southern Adriatic. *Journal of Marine Systems* 20:255–278.

Kovalev, A. V., A. E. Kideys, E. V. Pavlova, A. A. Shmeleva, V. A. Skryabin, N. A. Ostrovskaya, and Z. Uysal. 1999. Composition and abundance of zooplankton of the eastern Mediterranean Sea. In P. Malanotte-Rizzoli and V. N. Eremeev, eds., *The Eastern Mediterranean as a Laboratory Basin for the Assessment of Contrasting Ecosystems*, pp. 81–95. Dordrecht: Kluwer Academic.

Kowalewski, M. 2002. The fossil record of predation: An overview of analytical methods. *Paleontological Society Papers* 8:3–40.

Kowalewski, M., and J. H. Nebelsick. 2003. Predation on Recent and fossil echinoids. In P. H. Kelley, M. Kowalewski, and T. A. Hansen, eds., *Predator–Prey Interactions in the Fossil Record*, pp. 279–302. New York: Kluwer Academic/Plenum.

Kowalewski, M., A. Dulai, and F. T. Fürsich. 1998. A fossil record full of holes: The Phanerozoic history of drilling predation. *Geology* 26:1091–1094.

———. 1999. A fossil record full of holes: The Phanerozoic history of drilling predation. Reply. *Geology* 27:959–960.

Krajkar, V. 2003. Climatology of geostrophic currents in the northern Adriatic. *Geofizika* 20:105–114.

Kruse, S. E., and L. H. Royden. 1994. Bending and unbending of an elastic litho-sphere: The Cenozoic history of the Apennine and Dinaride foredeep basins. *Tectonics* 13:278–302.

Lagaaij, R., and Y. V. Gautier. 1965. Bryozoan assemblages from marine sedi-ments of the Rhône delta, France. *Micropaleontology* 11:39–58.

Lambeck, K., T. M. Esat, and E.-K. Potter. 2002. Links between climate and sea levels for the past three million years. *Nature* 419:199–206.

Lawrence, D. R. 1968. Taphonomy and information losses in fossil communities. *Geological Society of America Bulletin* 79:1315–1330.

Lee, C. M., F. Askari, J. Book, S. Carniel, B. Cushman-Roisin, C. Dorman, J. Doyle, P. Flament, C. K. Harris, B. H. Jones, M. Kuzmić, P. Martin, A. Ogs-ton, M. Orlić, H. Perkins, P.-M. Poulain, J. Pullen, A. Russo, C. Sherwood, R. P. Signeli, and D. Thaler Detweiler. 2005. Northern Adriatic response to a wintertime bora wind event. *EOS* 86:157, 163, 165.

Leighton, L. R. 2003. Predation on brachiopods. In P. H. Kelley, M. Kowalewski, and T. A. Hansen, eds., *Predator–Prey Interactions in the Fossil Record*, pp. 215–237. New York: Kluwer Academic/Plenum.

Lescinsky, H.,L. 1993. Taphonomy and paleoecology of epibionts on the scallops *Chlamys hastata* (Sowerby 1843) and *Chlamys rubida* (Hinds 1845). *Palaios* 8:267–277.

Lisac, I., B. Zelenko, A. Marki, and Ž. Trošić. 1998. Wind-direction frequency analysis for the Jugo wind in the Adriatic. *Hrvatski Meteoroloski Casopis* 33–34:19–37.

Lohrer, A. M., S. F. Thrush, and M. M. Gibbs. 2004. Bioturbators enhance ecosystem function through complex biogeochemical interactions. *Nature* 431:1092–1095.

Lukasik, J. J., N. P. James, B. McGowran, and Y. Bone. 2000. An epeiric ramp: Low-energy, cool-water carbonate facies in a Tertiary inland sea, Murray Ba-sin, South Australia. *Sedimentology* 47:851–881.

Malanotte-Rizzoli, P., and A. Bergamasco. 1983. The dynamics of the coastal region of the northern Adriatic Sea. *Journal of Physical Oceanography* 13: 1105–1130.

Malej, A. R. P., and R. P. Harris. 1993. Inhibition of copepod grazing by diatom exudates: A factor in the development of mucus aggregates? *Marine Ecology Progress Series* 98:33–42.

Malej, A., P. Mozetič, V. Malačič, S. Terzic, and M. Ahel. 1995. Phytoplankton re-sponses to freshwater inputs in a small semi-enclosed gulf (Gulf of Trieste, Adriatic Sea). *Marine Ecology Progress Series* 120:111–121.

Malinverno, A., and W. B. F. Ryan. 1986. Extension in the Tyrrhenian Sea and shortening in the Apennines as result of arc migration driven by sinking of the lithosphere. *Tectonics* 5:227–245.

Manca, B., and A. Giorgetti. 1999. Flow patterns of the main water masses across transversal areas in the southern Adriatic Sea: Seasonal variability.

In P. Malanotte-Rizzoli and V. N. Eremoov, eds., *The Eastern Mediterranean as a Laboratory Basin for Assessment of Contrasting Ecosystems*, pp. 495–506. Dordrecht: Kluwer Academic.

Manca, B. B., V. Kovačević, M. Gačić, and D. Viezzoli. 2002. Dense water formation in the southern Adriatic Sea and spreading into the Ionian Sea in the period 1997–1999. *Journal of Marine Systems* 33–34:133–154.

Mancinelli, G., S. Faze, and L. Rossi. 1998. Sediment structural properties mediating dominant feeding types patterns in soft-bottom macrobenthos of the northern Adriatic Sea. *Hydrobiologia* 367:211–222.

Marchini, A., K. Gauzer, and A. Occhipinti-Ambrogi. 2004. Spatial and temporal variability of hard-bottom macrofauna in a disturbed coastal lagoon (Sacca di Goro, Po River Delta, Northwestern Adriatic Sea). *Marine Pollution Bulletin* 48:1084–1095.

Mariucci, M. T., A. Amato, and P. Montone. 1999. Recent tectonic evolution and present stress in the Northern Apennines (Italy). *Tectonics* 18:108–118.

Martin, R. E. 1996. Secular increase in nutrient levels through the Phanerozoic: Implications for productivity, biomass, and diversity of the marine biosphere. *Palaios* 11:209–219.

Márton, E., D. Pavelić, B. Tomljenovic, R. Avanić, J. Pamić and P. Márton. 2002a. In the wake of a counterclockwise rotating Adriatic microplate: Neogene paleomagnetic results from northern Croatia. *International Journal of Earth Sciences (Geologisches Rundschu)* 91:514–523.

Márton, E., L. Fodor, B. Jelen, P. Marton, H. Rifelj, and R. Kevrić. 2002b. Miocene to Quaternary deformation in NE Slovenia: Complex paleomagnetic and structural study. *Journal of Geodynamics* 34:627–651.

Márton, E., K. Drobne, V. Ćosović, and A. Moro. 2003. Palaeomagnetic evidence for Tertiary counterclockwise rotation of Adria. *Tectonophysics* 377:143–156.

Massari, F., P. Grandesso, C. Stefani, and P. G. Jobstraibizer. 1986. A small polyhistory foreland basin evolving in a context of oblique convergence: The Venetian basin (Chattian to Recent, Southern Alps, Italy). *International Association of Sedimentologists Special Publication* 8:141–168.

Mattei, N., and M. Pellizzato. 1996. A population study on three stocks of a commercial Adriatic pectinid (*Pecten jacobaeus*). *Fisheries Research* 26:49–65.

Mauri, E., and P.-M. Poulain. 2001. Northern Adriatic Sea surface circulation and temperature/pigment fields in September and October 1997. *Journal of Marine Systems* 29:51–87.

McKerrow, W. S., ed. 1978. *The Ecology of Fossils.* Cambridge, MA: MIT Press.

McKinney, F. K. 1993. A faster-paced world?: Contrasts in biovolume and life-process rates in cyclostome (Class Stenolaemata) and cheilostome (Class Gymnolaemata) bryozoans. *Paleobiology* 19:335–351.

———. 1996. Encrusting organisms on co-occurring disarticulated valves of two marine bivalves: Comparison of living assemblages and skeletal residues. *Paleobiology* 22:543–567.

———. 2003. Preservation potential and paleoecological significance of epibenthic suspension feeder-dominated benthic communities (northern Adriatic Sea). *Palaios* 18:47–62.

McKinney, F. K., and S. J. Hageman. 2006. Paleozoic to modern marine ecological shift displayed in the northern Adriatic Sea. *Geology* 34:881–884.

McKinney, F. K., and J. B. C. Jackson. 1989. *Bryozoan Evolution*. Boston: Unwin Hyman.

McKinney, F. K., and A. Jaklin. 1993. Living populations of free-lying bryozoans: Implications for post-Paleozoic decline of the growth habit. *Lethaia* 26:171–179.

———. 2000. Spatial niche partitioning in the *Cellaria* meadow epibiont association, northern Adriatic Sea. *Cahiers de Biologie Marine* 41:1–17.

———. 2001. Sediment accumulation in a shallow-water meadow carpeted by a small erect bryozoan. *Sedimentary Geology* 145:397–410.

McKinney, F. K., and M. J. McKinney. 1991. *Exercises in Invertebrate Paleontology*. Boston: Blackwell Scientific.

McKinney, F. K., and P. D. Taylor. 2003. Palaeoecology of free-lying domal bryozoan colonies from the Upper Eocene of southeastern USA. *Acta Palaeontologica Polonica* 48:447–462.

McKinney, F. K., P. D. Taylor, and S. Lidgard. 2003. Predation on bryozoans and its reflection in the fossil record. In P. H. Kelley, M. Kowalewski, and T. A. Hansen, eds., *Predator–Prey Interactions in the Fossil Record*, pp. 239–261. New York: Kluwer Academic/Plenum.

Mele, G. 2001. The Adriatic lithosphere is a promontory of the African plate: Evidence of a continuous mantle lid in the Ionian Sea from efficient *Sn* propagation. *Geophysical Research Letters* 28:431–434.

Michelato, A., and V. Kovačević. 1991. Some dynamic features of the flow through the Otranto Strait. *Bollettino di Oceanologie Teorica ed Applicata* 9:39–51.

Miljush, P. 1973. Geologic-tectonic structure and evolution of outer Dinarides and Adriatic area. *American Association of Petroleum Geologists Bulletin* 57:913–929.

Miller, A. I. 1988. Spatio-temporal transitions in Paleozoic Bivalvia: An analysis of North American fossil assemblages. *Historical Biology* 1:251–273.

———. 1989. Spatio-temporal transitions in Paleozoic Bivalvia: A field comparison of Upper Ordovician and upper Paleozoic bivalve-dominated fossil assemblages. *Historical Biology* 2:227–260.

Mioković, D. 1999. Promjene u sastavu microfitoplanktona sjevernog Jadrana od 1972.–1994. godine. In D. Gere, ed., *2. Hrvatska Konferencija o Vodama: "Hrvatske Vode od Jadrana do Dunava": Zbornik Radova, Dubrovnik 19–22. Svibnja 1999*, pp. 233–243. Zagreb: Hrvatske vode.

Monegatti, P., S. Raffi, and G. Raineri. 1997. The Monte Falcone–Rio Riorzo composite section: Biostratigraphic and ecobiostratigraphic remarks. *Bollettino della Società geologica Italiana* 36:245–260.

Montone, P., M. T. Mariucci, S. Pondrelli, and A. Amato. 2004. An improved stress map for Italy and surrounding regions. *Journal of Geophysical Research* 109:B10410.

Morello, E. B., C. Froglia, R. J. A. Atkinson, and P. G. Moore. 2005. Impacts of hydraulic dredging on a macrobenthic community of the Adriatic Sea, Italy. *Canadian Journal of Fisheries and Aquatic Sciences* 62:2076–2087.

Moruzzi, G., and U. Follador. 1973. Il miocene superiore ed il pliocene inferiore della zona dello Scoglio del Trave (tra Ancona ed il Monte Conero–Marche) eloro inquadramento geologico regionale. *Geologica Romana* 12:129–149.

Mosetti, F., and A. Lavenia. 1969. Ricerche oceanografiche in Adriatico nel periodo 1966–1968. *Bollettino de Geofisica Teorica ed Applicata* 11:191–218.

Müller, W. E. G., R. K. Zahn, B. Kurelec, and I. Müller. 1984. A catalogue of the sponges near Rovinj. *Thalassia Jugoslavica* 20:13–23.

Müller, W. E. G., S. Riemer, B. Kurelec, N. Smodlaka, S. Puškarić, B. Jagic, G. Müller-Niklas, and N. V. Queric. 1998. Chemosensitizers of the multixenobiotic resistance in amorphous aggregates (marine snow): Etiology of mass killing on the benthos in the northern Adriatic. *Environmental Toxicology and Pharmacology* 6:229–238.

Najdek, M., and D. Degobbis. 1997. The relative importance of autochthonous and allochthonous sources of organic matter in the northwestern Adriatic sediment. *Periodicum Biologorum* 99:181–191.

Najdek, M., D. Degobbis, D. Mioković, and I. Ivančić. 2002. Fatty acid and phytoplankton compositions of different types of mucilaginous aggregates in the northern Adriatic. *Journal of Plankton Research* 24:429–441.

Nebelsick, J. H., B. Schmid, and M. Stachowitsch. 1997. The encrustation of fossil and recent sea-urchin tests: Ecological and taphonomic significance. *Lethaia* 30:271–284.

Newell, N. D. 1959. The nature of the fossil record. *Proceedings of the American Philosophical Society* 103:65–75.

Newell, R. I. E., and J. A. Ott. 1999. Macrobenthic communities and eutrophication. In T. J. Malone, A. Malej, L. W. Harding Jr., N. Smodlaka, and R. E. Turner, eds., *Ecosystems at the Land–Sea Margin*, pp. 265–293. Washington: American Geophysical Union.

Newton, R. S., and A. Stefanon. 1975. The "Tegnue de Ciosa" area: Patch reefs in the northern Adriatic Sea. *Marine Geology* 19:M27–M33.

———. 1982. Side-scan sonar and subbottom profiling in the northern Adriatic Sea. *Marine Geology* 46:279–306.

Nicholson, H. A. 1872. *A Manual of Palaeontology.* Edinburgh: William Blackwood and Sons.

Nikolić, M. 1959. Doprinos poznavanju briozojskih asocijacija I. *Thalassia Jugoslavica* 1:69–80.

Nixon, S. W. 1995. Coastal marine eutrophication: A definition, social causes, and future concerns. *Ophelia* 41:199–219.

Novosel, M., and A. Požar-Domac. 2001. Checklist of Bryozoa of the eastern Adriatic Sea. *Natura Croatica* 10:387–421.

Novosel, M., G. Olujic, S. Cocito, and A. Požar-Domac. 2005. Submarine freshwater springs in the Adriatic Sea: A unique habitat for the bryozoan *Pentapora fascialis*. In H. I. Moyano, J. M. Cancino, and P. N. Wyse Jackson, *Bryozoan Studies 2004*, pp. 215–221. Leiden: A. A. Balkema.

Occhipinti-Ambrogi, A., D. Savini, and G. Forni. 2005. Macrobenthos community structural changes off Cesenatico coast (Emilia Romagna, Northern Adriatic): A six-year monitoring programme. *Science of the Total Environment* 353:317–328.

Oldfield, F., A. Asioli, C. A. Accorsi, A. M. Mercri, S. Juggins, L. Langone, T. Rolph, F. Trincardi, G. Wolff, Z. Gibbs, L. Vigliotti, M. Frignani, K. van der Pot, and N. Branch. 2003. A high-resolution Late Holocene palaeoenvironmental record from the central Adriatic. *Quaternary Science Reviews* 22:319–342.

Oldow, J. S., L. Ferranti, D. S. Lewis, J. K. Campbell, B. D'Argenio, R. Catalano, G. Pappone, L. Carmignani, P. Conti, and C. L. V. Aiken. 2002. Active fragmentation of Adria, the north African promontory, central Mediterranean orogen. *Geology* 30:779–782.

Ölscher, E. M., and K. Fedra. 1977. On the ecology of a suspension-feeding benthic community: Filter efficiency and behaviour. In B. F. Keegan, P. O. Ceidigh, and P. J. S. Boaden, eds., *Biology of Benthic Organisms*, pp. 483–492. Oxford: Pergamon.

Orel, G., and B. Mennea. 1969. I popolamenti bentonici di alcuni tipi di fondo mobile del Golfo di Trieste. *Pubblicazione Stazione Zoologica di Napoli* 37 (suppl.):261–276.

Orel, G., R. Marocco, E. Vio, D. Del Piero, and G. Della Seta. 1987. Sedimenti e biocenosi bentoniche tra La Foce del Po ed il Golfo di Trieste (Alto Adriatico). *Bulletin d'Ecologie* 18:229–241.

Ori, G. G., M. Rovieri, and F. Vannoni. 1986. Plio-Pleistocene sedimentation in the Apenninic-Adriatic foredeep (Central Adriatic Sea, Italy). *International Association of Sedimentologists Special Publication* 8:183–198.

Orlić, M. 1989. Salinity of the North Adriatic: A fresh look at some old data. *Bollettino di Oceanologia Teorica ed Applicata* 7:219–228.

Orlić, M., M. Kuzmić, and Z. Vučak. 1986. Wind-curl currents in the Northern Adriatic and formulation of bottom friction. *Oceanologica Acta* 9:425–431.

Orlić, M., M. Gačić, and P. E. La Violette. 1992. The currents and circulation of the Adriatic Sea. *Oceanologica Acta* 15:109–124.

Orlić, M., M. Kuzmić, and Z. Pasaric. 1994. Response of the Adriatic Sea to the bora and sirocco forcing. *Continental Shelf Research* 14:91–116.

Orth, R. J. 1975. Destruction of eelgrass, *Zostera marina*, by the cownose ray, *Rhinoptera bonasus*, in the Chesapeake Bay. *Chesapeake Science* 16:205–208.

Ott, J. A. 1992. The Adriatic benthos: Problems and perspectives. In G. Colombo, I. Ferrari, V. U. Ceccherelli, and R. Rossi, eds., *Marine Eutrophication and Population Dynamics*, pp. 367–378. Fredensborg, Denmark: Olsen and Olsen.

Paklar, G. B., V. Isakov, D. Doracin, V. Kourafalou, and M. Orlić. 2001. A case study of bora-driven flow and density changes on the Adriatic Shelf (January 1987). *Continental Shelf Research* 21:1751–1783.

Parisi, V., R. Ambrogi, D. Bedulli, M. G. Mezzadri, and P. Poli. 1985. Struttura e dinamica dei popolamenti bentonici negli ambienti sedimentari del Delta Padano. *Nova Thalassia* 7:215–251.

Pasini, G., M. L. Colalongo, P. V. Curzi, and M. Taviani. 1993. Analisi paleoecologica degli ambienti tardo-quaternari del Dosso Gallignani (Adriatico centro-meridionale) mediante lo studio paleontologico de carote. *Bollettino della Società paleontologica Italiana* 32:113–130.

Patacca, E., R. Sartori, and P. Scandone. 1990. Tyrrhenian basin and Apenninic arcs: Kinematic relations since Late Tortonian times. *Memorie della Società geologica Italiana* 45:425–451.

Patterson, C. 1993. Osteichthyes: Teleostei. In M. J. Benton, ed., *The Fossil Record 2*, pp. 621–663. London: Chapman and Hall.

Pavlovec, R., M. Pleničar, K. Drobne, B. Oborelec, and F. Šušteršič. 1987. History of geological investigations of the Karst (Kras) region and the neighboring territory (Western Dinarides). *Memorie della Società geologica Italiana* 40:9–20.

Pazold, J., H. Ristedt, and G. Wefer. 1987. Rate of growth and longevity of a large colony of *Pentapora foliacea* (Bryozoa) recorded in their oxygen isotope profiles. *Marine Biology* 96:535–538.

Peharda, M., C. A. Richardson, V. Onofri, A. Bratoš, and M. Crnčević. 2002. Age and growth of the bivalve *Arca noae* L. in the Croatian Adriatic Sea. *Journal of Molluscan Studies* 68:307–310.

Penna, N., S. Capellacci, and F. Ricci. 2004. The influence of the Po River discharge on phytoplankton bloom dynamics along the coastline of Pesaro (Italy) in the Adriatic Sea. *Marine Pollution Bulletin* 48:321–326.

Pérès, J. M., and J. Picard. 1964. Manuel de bionomie benthique de la Mer Mediteranée. *Recueil Travaux de la Station marine d'Endoume* 31(47):1–147.

Pettine, M., L. Patrolecco, M. Camusso, and S. Crescenzio. 1998. Transport of carbon and nitrogen to the northern Adriatic Sea by the Po River. *Estuarine, Coastal, and Shelf Science* 46:127–142.

Pettine, M., L. Patrolecco, M. Manganelli, S. Capri, and M. G. Farrace. 1999. Seasonal variations of dissolved organic matter in the northern Adriatic Sea. *Marine Chemistry* 64:153–169.

Phillips, J. 1840. Palaeozoic series. *Penny Cyclopaedia* 17:153–154.

———. 1841. *Figures and Descriptions of the Palaeozoic Fossils of Cornwall, Devon, and East Somerset*. London: Longman, Brown, Green, and Longmans.

Picha, F. J. 2002. Late orogenic strike-slip faulting and escape tectonics in frontal Dinarides-Hellenides, Croatia, Jugoslavia, Albania, and Greece. *AAPG Bulletin* 86:1659–1671.

Pigorini, B. 1968. Sources and dispersion of Recent sediments of the Adriatic Sea. *Marine Geology* 6:187–229.

Pirazzoli, P. A., and A. Tomasin. 2002. Recent evolution of surge-related events in the northern Adriatic area. *Journal of Coastal Research* 18:537–554.

Pisarović, A., V. Ž. Meixner, and S. Benc. 2000. A contribution to the knowledge of bivalve *Corbula gibba* (Olivi, 1792) behaviour, oxygen consumption, and anaerobic metabolism. *Periodicum Biologorum* 102:303–307.

Pollak, M. J. 1951. The sources of the deep water of the eastern Mediterranean Sea. *Journal of Marine Research* 10:128–152.

Poluzzi, A. 1975. I briozoi cheilostomi del Pliocene della Val d'Arda (Piacenza, Italia). *Memorie della Società Italiana di Scienze Naturali Museo Civico di Storia Naturale di Milano* 21:37–77.

———. 1979. I briozoi dei fondi mobili tra Ancona e la Foce del Fiume Reno (Adriatico settentrionale). *Giornale di Geologia* 43:237–256.

Poluzzi, A., and A. Agnoletto. 1988. I briozoi cheilostomi incrostanti come indicatori ambientali in aree di delta: Un'applicazione dell'analisi discriminante. *Acta Naturalia de L'Ateno Parmense* 24:1–17.

Poluzzi, A., and P. Forti. 1983. Substrati, distribuzione areale e interazione competitive di alcuni briozoi cheilostomi della piana deltizia del Fiume Po. *Bolletino della Società paleontologica Italiana* 22:53–64.

Poluzzi, A., and M. Taviani. 1988. Fossilizzazione de faune bentoniche in un'area de delta (Fiume Po, Italia del Nord). *Quarto Simposio de Ecologie e Paleoecologia delle Comunità Bentoniche*: 731–752.

Poluzzi, A., P. V. Curzi, and M. Baldani. 1985. Analisi quantitative di facies sedimentarie ed ambienti deposizionali dell'Adriatico centrale. *Acta-Naturalia de "L'Ataneo Permense"* 21:59–78.

Poluzzi, A., R. Capozzi, G. Giordani, and M. Venturini. 1988. I briozoi dello spungone nel terreni pliocenici della Romagne. *Acta-Naturalia de "L'Ataneo Permense"* 24:19–82.

Pompei, M., C. Mazziotti, F. Guerrini, M. Cangini, S. Pigozzi, M. Benzi, S. Palamidesi, L. Boni, and R. Pistocchi. 2003. Correlation between the presence of *Gonyaulax fragilis* (Dinophyceae) and the mucilage phenomena of the Emilia-Romagna coast (northern Adriatic Sea). *Harmful Algae* 2: 301–316.

Poulain, P.-M. 1999. Drifter observations of surface circulation in the Adriatic Sea between December 1994 and March 1996. *Journal of Marine Systems* 20:231–253.

———. 2001. Adriatic Sea surface circulation as derived from drifter data between 1990 and 1999. *Journal of Marine Systems* 29:3–32.

Poulain, P.-M., and F. Raicich. 2001. Forcings. In B. Cushman-Roisin, M. Gačić, P.-M. Poulain, and A. Artegiani, eds., *Physical Oceanography of the Adriatic Sea*, pp. 45–65. Dordrecht: Kluwer Academic.

Poulain, P.-M., M. Gačić, and A. Vetrano. 1996. Current measurements in the Strait of Otranto reveal unforseen aspects of its hydrodynamics. *EOS* 77(36): 345, 348.

Poulain, P.-M., E. Mauri, and L. Ursella. 2004. Unusual upwelling event and current reversal off the Italian Adriatic coast in summer 2003. *Geophysical Research Letters* 31:L05303.

Pranovi, F., S. Raicevich, G. Franceschini, M.G. Farrace, and O. Giovanardi. 2000. Rapido trawling in the northern Adriatic Sea: Effects on benthic communities in an experimental area. *ICES Journal of Marine Science* 57:517–524.

Precali, R., M. Giani, M. Marini, F. Grilli, C.R. Ferrari, O. Pečar, and E. Paschini. 2005. Mucilaginous aggregates in the northern Adriatic in the period 1999–2002: Typology and distribution. *Science of the Total Environment* 353:10–23.

Prelogović, E., B. Aljinović, and S. Bahun. 1995. New data on structural relationships in the North Dalmatian Dinaride area. *Geologica Croatica* 48:167–176.

Puddu, A., R. La Ferla, A. Allegra, C. Bacci, M. Lopez, F. Oliva, and C. Pierotti. 1998. Seasonal and spatial distribution of bacterial production and biomass along a salinity gradient (northern Adriatic Sea). *Hydrobiologia* 363:271–282.

Puškarić, S., G.W. Berger, and F.J. Jorissen. 1990. Successive appearance of subfossil phytoplankton species in Holocene sediments of the northern Adriatic and its relation to the increased eutrophication pressure. *Estuarine, Coastal, and Shelf Science* 31:177–187.

Rabbi, E., and F. Ricci Lucchi. 1966. Stratigrafia e sedimentologia del Messiniano forlivese (dintorni de Predappio). *Giornale di Geologia* 34:593–638.

Radenbaugh, T.A., and F.K. McKinney. 1998. Comparison of the structure of a Mississippian and a Holocene pen shell assemblage. *Palaios* 13:52–69.

Raicich, F. 1994. Note on the flow rates of the Adriatic rivers. *Consiglio Nazionale delle Ricerche Istituto Sperimentale Talassografico Technical Report RF 02/94*, 8 pp.

———. 1996. On the fresh water balance of the Adriatic Sea. *Journal of Marine Systems* 9:305–319.

Rasmussen, K.A., and C.E. Brett. 1985. Taphonomy of Holocene cryptic biotas from St. Croix, Virgin Islands: Information loss and preservational biases. *Geology* 13:551–553.

Raup, D.M., and J.J. Sepkoski Jr. 1982. Mass extinctions in the marine fossil record. *Science* 215:1501–1503.

Redfield, A.C., B.H. Ketchum, and F.A. Richards. 1963. The influence of organisms on the composition of sea-water. In M.N. Hill, ed., *The Sea: Ideas and Observations*, vol. 2, pp. 26–77. New York: Interscience.

Relini, G., F. Tixi, M. Relini, and G. Torchia. 1998. The macrofouling on offshore platforms at Ravenna. *International Biodeterioration and Biodegradation* 41:41–55.

Revelante, N., and M. Gilmartin. 1976. The effect of Po River discharge on phytoplankton dynamics in the northern Adriatic Sea. *Marine Biology* 34:259–271.

———. 1983. The phytoplankton of the Adriatic Sea: Community structure and characteristics. *Thalassia Jugoslavica* 19:303–318.

———. 1992. The lateral advection of particulate organic matter from the Po Delta region during summer stratification, and its implications for the northern Adriatic. *Estuarine, Coastal, and Shelf Science* 35:191–212.

Revelante, N., M. Gilmartin, and N. Smodlaka. 1985. The effects of Po River induced eutrophication on the distribution and community structure of ciliated protozoan and micrometazoan populations in the northern Adriatic Sea. *Journal of Plankton Research* 7:461–471.

Rhoads, D. C., and D. K. Young. 1970. The influence of deposit-feeding organisms on sediment stability and community trophic structure. *Journal of Marine Research* 28:150–178.

Ricci Lucchi, F. 1975. Miocene paleogeography and basin analysis in the Periadriatic Apennines. In C. H. Squyres, ed., *Geology of Italy*, pp. 129–236. Tripoli: Earth Sciences Society of the Libyan Arab Republic.

———. 1986. The Oligocene to Recent foreland basins of the northern Apennines. *International Association of Sedimentologists Special Publication* 8:105–139.

Richards, F. A. 1958. Dissolved silicate and related properties of some western North Atlantic and Caribbean waters. *Journal of Marine Research* 17: 449–465.

Richardson, C. A., M. Peharda, H. Kennedy, P. Kennedy, and V. Onofri. 2004. Age, growth rate, and season of recruitment of *Pinna nobilis* (L) in the Croatian Adriatic determined from Mg:Ca and Sr:Ca shell profiles. *Journal of Experimental Marine Biology and Ecology* 299:1–16.

Ridente, D., and F. Trincardi. 2002. Eustatic and tectonic control on deposition and lateral variability of Quaternary regressive sequences in the Adriatic basin (Italy). *Marine Geology* 184:273–293.

———. 2005. Pleistocene "muddy" forced-regression deposits on the Adriatic shelf: A comparison with prodelta deposits of the late Holocene highstand mud wedge. *Marine Geology* 222–223:213–233.

Rizzini, A., and L. Dondi. 1980. Messinian evolution of the Po Basin and its economic implications (hydrocarbons). *Palaeogeography, Palaeoclimatology, Palaeoecology* 29:41–74.

Robertson, A. H. F., and M. Grasso. 1995. Overview of the Late Tertiary–Recent tectonic and palaeo-environmental development of the Mediterranean region. *Terra Nova* 7:114–127.

Rhodes, M. C. and R. J. Thompson. 1993. Comparative physiology of suspension-feeding in living brachiopods and bivalves; evolutionary implications. *Paleobiology* 19:322–334.

Roether, W., and R. Schlitzer. 1991. Eastern Mediterranean deepwater renewal on the basis of chlorofluoromethane and tritium data. *Dynamics of Atmospheres and Oceans* 15:333–354.

Rosenbaum, G., G. S. Lister, and C. Duboz. 2004. The Mesozoic and Cenozoic motion of Adria (central Mediterranean): A review of constraints and limitations. *Geodinamica Acta* 17:125–139.

Roure, F., P. Casero, and R. Vially. 1991. Growth processes and melange formation in the southern Apennines accretionary wedge. *Earth and Planetary Science Letters* 102:395–413.

Rudwick, M. J. S. 1972. *The Meaning of Fossils*. London: MacDonald.

———. 1985. *The Great Devonian Controversy*. Chicago: University of Chicago Press.

Sala, E. 1997. Fish predators and scavengers of the sea urchin *Paracentrotus lividus* in protected areas of the north-west Mediterranean Sea. *Marine Biology* 129:531–539.

Sartoretto, S., M. Verlaque, and J. Laborel. 1996. Age of settlement and accumulation rate of submarine "coralligène" (−10 to −60 m) of the northwestern Mediterranean Sea: Relation to Holocene rise in sea level. *Marine Geology* 130:317–331.

Sathyendranath, S., A. Longhurst, C. M. Caverhill, and T. Platt. 1995. Regionally and seasonally differentiated primary production in the North Atlantic. *Deep-Sea Research* 142:1773–1802.

Savrda, C. E., and D. L. Bottjer. 1991. Oxygen-related biofacies in marine strata: An overview and update. In R. V. Tyson and T. H. Pearson, eds., *Modern and Ancient Continental Shelf Anoxia* (Special Publication 58), pp. 201–219. London: Geological Society.

Scardi, M., R. Crema, P. Di Dato, E. Fresi, and G. Orel. 2000. Le comunità bentoniche dell'alto Adriatico: un'analisi preliminare dei cambiamenti strutturali dagli anni '30 ad oggi. In Giovanardi, O., ed., *Impact of trawl fishing on benthic communities — Proceedings*, pp. 95–108. Rome: Istituto Centrale per la Ricerca Scientifica e Tecnologica Applicata al Mare.

Scarponi, D., and M. Kowalewski. 2004. Stratigraphic paleoecology: Bathymetric signatures and sequence overprint of mollusk associations from upper Quaternary sequences of the Po Plain, Italy. *Geology* 32:989–992.

Schäfer, W. 1972. *Ecology and Paleoecology of Marine Environments*. Chicago: University of Chicago Press.

Schaumberg, G. 1979. Neue nachweise von Bryozoen und Brachiopoden Nahrung des permischen Holocephalen, *Janass bitumminosa* (Schlotheim). *Philippia* 4:3–11.

Schettino, A., and C. Scotese. 2002. Global kinematic constraints to the tectonic history of the Mediterranean region and surrounding areas during the Jurassic and Cretaceous. *Journal of the Virtual Explorer* 8:149–168.

Schinner, F., M. Stachowitsch, and H. Hilgers. 1997. Loss of benthic communities: Warning signal for coastal ecosystem management. *Aquatic Conservation: Marine and Freshwater Ecosystems* 6:343–352.

Schinner, G. O. 1993. Burrowing behavior, substratum preference, and distribution of *Schizaster canaliferus* (Echinoidea: Spatangoida) in the northern Adriatic Sea. *Marine Ecology* 14:129–145.

Schopf, T. J. M. 1978. Fossilization potential of an intertidal fauna: Friday Harbor, Washington. *Paleobiology* 4:261–270.

Secord, J. A. 1986. *Controversy in Victorian Geology*. Princeton: Princeton University Press.

Sellner, K. G., and S. Fonda Umani. 1999. Dinoflagellate blooms and mucilage production. In T. C. Malone, A. Malej, L. W. Harding Jr., N. Smodlaka, and R. E. Turner, eds., *Ecosystems at the Land–Sea Margin*, pp. 173–206. Washington: American Geophysical Union.

Seneš, J. 1988a. Principles of study of Adriatic shelf ecosystems from the viewpoint of applications in geology. *Geologicky Zborník–Geologica Carpathica* 39:385–300.

———. 1988b. Quantitative analysis of North and South Adriatic shelf ecosystems. *Geologicky Zborník–Geologica Carpathica* 39:675–712.

———. 1988c. The Island Banjole: A type region of recent marine ecosystems on North Adriatic shelf. *Geologicky Zborník–Geologica Carpathica* 39:713–738.

———. 1989. North Adriatic inter-island shelf ecosystems of the Rovinj area. *Geologicky Zborník–Geologica Carpathica* 40:333–354.

———. 1990. Some infra- and circalittoral ecosystems of the eastern part of the South Adriatic shelf. *Geologicky Zborník–Geologica Carpathica* 41:199–228.

Sepkoski, J. J., Jr. 1978. A kinetic model of Phanerozoic taxonomic diversity, I: Analysis of marine orders. *Paleobiology* 4:223–251.

———. 1979. A kinetic model of Phanerozoic taxonomic diversity, II: Early Phanerozoic families and multiple equilibria. *Paleobiology* 5:222–251.

———. 1981. A factor analytic description of the Phanerozoic marine fossil record. *Paleobiology* 7:36–53.

———. 1982. A compendium of fossil marine families. *Milwaukee Public Museum Contributions in Biology and Geology* 51:1–125.

———. 1984. A kinetic model of Phanerozoic taxonomic diversity, III: Post-Paleozoic families and mass extinctions. *Paleobiology* 10:246–267.

Sepkoski, J. J., Jr., and A. I. Miller. 1985. Evolutionary faunas and the distribution of Paleozoic marine communities in space and time. In J. W. Valentine, ed., *Phanerozoic Diversity Patterns*, pp. 153–190. Princeton: Princeton University Press.

Sheehan, P. M., and D. R. J. Schiefelbein. 1984. The trace fossil *Thalassinoides* from the Upper Ordovician of the eastern Great Basin: Deep burrowing in the early Paleozoic. *Journal of Paleontology* 58:440–447.

Shennan, I., and P. L. Woodworth. 1992. A comparison of late Holocene and twentieth-century sea-level trends from the UK and North Sea region. *Geophysical Journal International* 109:96–105.

Signor, P. W., III, and C. E. Brett. 1984. The mid-Paleozoic precursor to the Mesozoic marine revolution. *Paleobiology* 10:229–245.

Simeoni, U., and M. Bondesan. 1997. The role and responsibility of man in the evolution of the Italian Adriatic coast. *Bulletin de l'Institut Océanographique, Monaco, Special Issue* 18:111–132.

Simeoni, U., N. Pano, and P. Ciavola. 1997. The coastline of Albania: Morphology, evolution, and coastal management issues. *Bulletin de l'Institut Océanographique, Monaco, Special Issue* 18:151–168.

Simonini, R., I. Annsaloni, A. M. Bonvicini Pagliai, and D. Prevedelli. 2004. Organic enrichment and structure of the macrozoobenthic community in the northern Adriatic Sea in an area facing Adige and Po mouths. *Journal of Marine Science* 61:871–881.

Sköld, M., and R. Rosenberg. 1996. Arm regeneration frequency in eight species of Ophiuroidea (Echinodermata) from European sea areas. *Journal of Sea Research* 35:353–362.

Sloan, N. A., and S. M. C. Robinson. 1983. Winter feeding by asteroids on a subtidal sandbed in British Columbia. *Ophelia* 22:125–140.

Smith, A. G. 1971. Alpine deformation and the oceanic areas of the Tethys, Mediterranean, and Atlantic. *Geological Society of America Bulletin* 82:2039–2070.

Smodlaka, N. 1986. Primary production of the organic matter as an indicator of the eutrophication in the northern Adriatic Sea. *Science of the Total Environment* 56:211–220.

Sondi, I., M. Juračić, and V. Pravdić. 1995. Sedimentation in a disequilibrium river-dominated estuary: The Raša River Estuary (Adriatic Sea, Croatia). *Sedimentology* 42:769–782.

Stachowitsch, M. 1980. The epibiotic and endolithic species associated with the gastropod shells inhabited by the hermit crabs *Paguristes oculatus* and *Pagurus cuanensis*. *Marine Ecology* 1:73–101.

———. 1984. Mass mortality in the Gulf of Trieste: The course of community destruction. *Marine Ecology (Pubblicazione della Statione zoologica di Napoli)* 5:243–264.

———. 1991. Anoxia in the northern Adriatic Sea: Rapid death, slow recovery. In R. V. Tyson and T. H. Pearson, eds., *Modern and Ancient Continental Shelf Anoxia* (Special Publication 58), pp. 119–129. London: Geological Society.

———. 1992. Benthic communities: Eutrophication's "memory mode." *Science of the Total Environment Supplement* 1992:1017–1028.

Stachowitsch, M., N. Fanuko, and M. Richter. 1990. Mucus aggregates in the Adriatic Sea: An overview of stages and occurrences. *Marine Ecology* 11:327–350.

Stanley, J. G. 1985. Species profiles: Life histories and environmental requirements of coastal fishes and invertebrates (mid-Atlantic)—hard clam. *U.S. Fish and Wildlife Service Biological Report* 82(11.41):1–24.

Stanton, R. J., Jr. 1976. Relationship of fossil communities to original communities of living organisms. In R. W. Scott and R. R. West, eds., *Structure and Classification of Paleocommunities*, pp. 107–129. Stroudsburg, PA: Dowden, Hutchinson, and Ross.

Starmans, A., J. Gutt, and W. E. Arntz. 1999. Mega-epibenthic communities in Arctic and Antarctic shelf areas. *Marine Biology* 135:169–280.

Stefani, M., and S. Vincenzi. 2005. The interplay of eustasy, climate, and human activity in the late Quaternary depositional evolution and sedimentary architecture of the Po Delta system. *Marine Geology* 222–223:19–48.

Stefanon, A. 1984. Sedimentologia del mare Adriatica: Rapporti tra erosione e sedimentazione olocenica. *Bollettino di Oceanologia Teorica ed Applicata* 2:281–324.

Števčić, Z. 1991. Decapod fauna of sea grass beds in the Rovinj area. *Acta Adriatica* 32:637–653.

Stravisi, F. 1983. Some characteristics of the circulation in the Gulf of Trieste. *Thalassia Jugoslavica* 19:343–349.

Sturm, B., M. Kuzmić, and M. Orlić. 1992. An evaluation and interpretation of CZCS-derived patterns on the Adriatic shelf. *Oceanologica Acta* 15:13–23.

Supić, N., and M. Orlić. 1999. Seasonal and interannual variability of the northern Adriatic surface fluxes. *Journal of Marine Systems* 20:205–229.

Supić, N., M. Orlić, and D. Degobbis. 2000. Istrian coastal countercurrent and its year-to-year variability. *Estuarine, Coastal, and Shelf Science* 51:385–397.

Taviani, M., M. Roveri, R. Impiccini, and L. Vigliotti. 1998 (1997). Segnalazione di Quaternario marino nella Val Chero (Appennino Piacentino). *Bollettino della Società paleontologica Italiana* 36:331–338.

Taylor, P. D. 1990. Bioimmured ctenostomes from the Jurassic and the origin of cheilostome Bryozoa. *Palaeontology* 33:19–34.

Taylor, P. D., and P. A. Allison. 1998. Bryozoan carbonates through time and space. *Geology* 26:459–462.

Tevesz, M. J. S., and P. L. McCall, eds. 1983. *Biotic Interactions in Recent and Fossil Benthic Communities.* New York: Plenum.

Thayer, C. W. 1979. Biological bulldozers and the evolution of marine benthic communities. *Science* 203:458–461.

———. 1983. Sediment-mediated biological disturbance and the evolution of marine benthos. In M. J. S. Tevesz and P. L. McCall, eds., *Biotic Interactions in Recent and Fossil Benthic Communities*, pp. 479–625. New York: Plenum.

Thayer, C. W. 1989. Recent articulate brachiopods; the effects of predation and competition for space. *Geological Society of America Abstracts with Programs* 12(2):86.

Thayer, C. W., J. Hall, K. Grage, and P. Barrett. 1992. Oligotrophic refuge for living brachiopods: The Southern Hemisphere and fiords of New Zealand. *Geological Society of America Abstracts with Programs* 24(7):313.

Tomadin, L. 2000. Sedimentary fluxes and different dispersion mechanisms of the clay sediments in the Adriatic Basin. *Rendiconti Fisiche Accademia dei Lincei* 9(11):161–174.

Tracey, S., J. A. Todd, and D. H. Erwin. 1993. Gastropoda. In M. J. Benton, ed., *The Fossil Record*, vol. 2, pp. 131–169. London: Chapman and Hall.

Trincardi, F., and A. Correggiari. 2000. Quaternary forced regression deposits in the Adriatic basin and the record of composite sea-level cycles. In D. Hunt and R. I. Gawthorpe, eds., *Sedimentary Responses to Forced Regressions* (Special Publications 172), pp. 245–269. London: Geological Society.

Trincardi, F., A. Correggiari, and M. Roveri. 1994. Late Quaternary transgressive erosion and deposition in a modern epicontinental shelf: The Adriatic semienclosed basin. *Geo-Marine Letters* 14:41–51.

Trincardi, F., A. Cattaneo, A. Asioli, A. Correggiari, and L. Langone. 1996. Stratigraphy of the late-Quaternary deposits in the central Adriatic basin and the record of short-term climatic events. *Memorie dell'Istituto Italiano di Idrobiologia* 55:39–70.

Trincardi, F., A. Cattaneo, A. Correggiari, and D. Ridente. 2004. Evidence of soft sediment deformation, fluid escape, sediment failure, and regional weak layers within the later Quaternary mud deposits of the Adriatic Sea. *Marine Geology* 213:91–119.

Twitchell, R. J., J. M. Feinberg, D. D. O'Connor, W. Alvarez, and L. B. McCollum. 2005. Early Triassic ophiuroids: Their paleoecology, taphonomy, and distribution. *Palaios* 20:213–223.

Udhayakumar, M., and A. A. Karande. 1989. Growth and breeding in cheilostome biofouler, *Electra benegalensis* Stoliczka, in Bombay waters, west coast of India. *India Journal of Marine Sciences* 18:95–99.

Udias, A. 1985. Seismicity of the Mediterranean Basin. In D. J. Stanley and F.-C. Wetzel, eds., *Geological Evolution of the Mediterranean Basin*, pp. 55–63. New York: Springer-Verlag.

Ulrich, E. O. 1890. Palaeozoic Bryozoa. *Illinois Geological Survey* 8: 283–688.

Urbani, R., E. Magaletti, P. Sist, and A. M. Cicero. 2005. Extracellular carbohydrates released by the marine diatoms *Cylindrotheca closterium*, *Thalassiosira pseudonana*, and *Skeletonema costatum*: Effect of P-depletion and growth status. *Science of the Total Environment* 353:300–306.

Valentine, J. W. 1989. How good was the fossil record? Clues from the Californian Pleistocene. *Paleobiology* 15:83–94.

Van Hoey, G., S. Degraer, and M. Vincx. 2004. Macrobenthic community structure of soft-bottom sediments at the Belgian Continental Shelf. *Estuarine, Coastal, and Shelf Science* 59:599–613.

van Straaten, L. M. J. U. 1971. Holocene and late-Pleistocene sedimentation in the Adriatic Sea. *Geologische Rundschau* 60:106–131.

Vatova, A. 1935. Ricerche preliminari sulle biocenosi del Golfo di Rovigno. *Thalassia* 2:1–30.

———. 1949. La fauna bentonica dell'alto e medio Adriatico. *Nova Thalassia* 1(3):1–110.

Venzo, G. A., and S. Stefanini. 1967. Distribuzione dei carbonati nei sedimenti di spiaggia e marini dell'Adriatico settentrionale tra Venezia e Trieste. *Studi Trentini di Scienze naturali (A)* 44:178–201.

Vermeij, G. J. 1977. The Mesozoic marine revolution: Evidence from snails, predators, and grazers. *Paleobiology* 3:245–258.

————. 1983. Shell-breaking predation through time. In M. J. S. Tevesz and P. M. McCall, eds., Biotic Interactions in Recent and Fossil Benthic Communities, pp. 649–669. New York: Plenum.

————. 1987. *Evolution and Escalation.* Princeton: Princeton University Press.

————. 1995. Economics, volcanoes, and Phanerozoic revolutions. *Paleobiology* 21:125–152.

Vezzani, L., and F. Ghisetti. 1998. *Carta geologica dell'Abruzzo,* Firenze, foglio est. Florence: S.E.L.C.A.

Vilibić, I., and M. Orlić. 2002. Adriatic water masses, their rates of formation and transport through the Otranto Strait. *Deep-Sea Research I* 49:1321–1340.

Vilibić, I., and N. Supić. 2003. Dense-water generation episodes in the northern Adriatic. *Il Nuovo Cimento* 27C:47–57.

————. 2005. Dense-water generation on a shelf: The case of the Adriatic Sea. *Ocean Dynamics* 55:403–415.

Vilibić, I., B. Grbec, and N. Supić. 2004. Dense water generation in the north Adriatic in 1999 and its recirculation along the Jabuka Pit. *Deep-Sea Research I* 51:1457–1474.

Vollenweider, R. A., F. Giovanardi, G. Montanari, and A. Rinaldi. 1998. Characterization of the trophic conditions of marine coastal waters with special reference to the NW Adriatic Sea: Proposal for a trophic scale, turbidity, and generalized water quality index. *Environmetrics* 9:329–357.

Warner, G. F. 1977. *The Biology of Crabs.* New York: Van Nostrand Reinhold.

Wilson, W. H. 1991. Competition and predation in marine soft-sediment communities. *Annual Review of Ecology and Systematics* 21:221–241.

Winchester, S. 2001. *The Map That Changed the World.* London: Viking.

Wu, P., and K. Haines. 1998. Modeling the dispersal of Levantine Intermediate Water and its role in Mediterranean deep water formation. *Journal of Geophysical Research* 101:6591–6607.

Wurzian, R. S. 1977. Predator–prey interaction between the crab *Pilumnus hirtellus* (Leach) and the brittle star *Ophiothrix quinquemaculata* (D. Chiaje) on a mutual sponge substrate. In B. F. Keegan, P. O. Ceidigh, and P. J. S. Boaden, eds., *Biology of Benthic Organisms,* pp. 613–620. Oxford: Pergamon.

Zabala, M., and P. Maluquer. 1988. Illustrated keys for the classification of Mediterranean Bryozoa. *Treballs del Museu de Zoologia Barcelona* 4:1–294.

Zahtila, E. 1997. Offshore polychaete fauna in the northern Adriatic with trophic characteristic. *Periodicum Biologorum* 99:213–217.

Zavatarelli, M., J. W. Baretta, J. G. Baretta-Bekka, and N. Pinardi. 2000. The dynamics of the Adriatic Sea ecosystem: An idealized model study. *Deep-Sea Research I* 47:937–970.

Zavatarelli, M., F. Raicich, D. Bregant, A. Russo, and A. Artegiani. 1998. Climatological biogeochemical characteristics of the Adriatic Sea. *Journal of Marine Systems* 18:227–263.

Zavodnik, D. 1967. Dinamika litoralnega fitala na Zahodnoistrski obali (Dynamics of the littoral phytal on the west coast of Istria). *Razprave–Dissertationes IV Razred Slovenska Akademija Znanosti in Umetnosti Academia Scientiarum et Artium Slovenica* 10:1–67.

————. 1971. Contribution to the dynamics of benthic communities in the region of Rovinj (northern Adriatic). *Thalassia Jugoslavica* 7:447–514.

————. 1980. Distribution of Echinodermata in the North Adriatic insular region. *Acta Adriatica* 21:437–468.

————. 1983. 400 years of the Adriatic marine science. *Thalassia Jugoslavica* 19:405–429.

Zavodnik, D., M. Hrs-Brenko, and M. Legac. 1991. Synopsis on the fan shell *Pinna nobilis* L. in the eastern Adriatic Sea. In C. F. Boudouresque, M. Avon, and V. Gravez, eds., *Les espèces marines à protéger in Méditerranée*, pp. 169–178. Marseille: GIS Posidonie.

Zoppini, A., M. Pettine, C. Totti, A. Puddu, A. Artegiani, and R. Pagnotta. 1995. Nutrients, standing crop, and primary production in western coastal waters of the Adriatic Sea. *Estuarine, Coastal, and Shelf Science* 41:493–513.

Zorè, M. 1956. On gradient currents in the Adriatic Sea. *Acta Adriatica* 8(6):1–38.

Zore-Armanda, M. 1983. Some physical characteristics of the Adriatic Sea. *Thalassia Jugoslavica* 19:433–450.

Zore-Armanda, M., and M. Gačić. 1987. Effects of bura on the circulation in the North Adriatic. *Annales Geophysicae* 5B:93–102.

Zuschin, M., and P. Perversler. 1996. Secondary hardground-communities in the northern Gulf of Trieste, Adriatic Sea. *Senckenbergiana Maritima* 28:53–63.

Zuschin, M., and W. E. Piller. 1994. Sedimentology and facies zonation along a transect through the Gulf of Trieste, northern Adriatic Sea. *Beiträge zur Paläontologie* 18:75–114.

Zuschin, M., M. Stachowitsch, P. Pervesler, and H. Killmann. 1999. Structural features and taphonomic pathways of a high-biomass epifauna in the northern Gulf of Trieste, Adriatic Sea. *Lethaia* 32:299–317.